Workplace Safety and Health

Assessing Current Practices
and Promoting Change
in the Profession

Occupational Safety and Health Guide Series

Series Editor

Thomas D. Schneid
Eastern Kentucky University
Richmond, Kentucky

Published Titles

The Comprehensive Handbook of School Safety, *E. Scott Dunlap*

Corporate Safety Compliance: OSHA, Ethics, and the Law, *Thomas D. Schneid*

Creative Safety Solutions, *Thomas D. Schneid*

Disaster Management and Preparedness, *Thomas D. Schneid and Larry R. Collins*

Discrimination Law Issues for the Safety Professional, *Thomas D. Schneid*

Labor and Employment Issues for the Safety Professional, *Thomas D. Schneid*

Loss Control Auditing: A Guide for Conducting Fire, Safety, and Security Audits, *E. Scott Dunlap*

Loss Prevention and Safety Control: Terms and Definitions, *Dennis P. Nolan*

Managing Workers' Compensation: A Guide to Injury Reduction and Effective Claim Management, *Keith R. Wertz and James J. Bryant*

Motor Carrier Safety: A Guide to Regulatory Compliance, *E. Scott Dunlap*

Occupational Health Guide to Violence in the Workplace, *Thomas D. Schneid*

Physical Hazards of the Workplace, *Larry R. Collins and Thomas D. Schneid*

Safety Performance in a Lean Environment: A Guide to Building Safety into a Process, *Paul F. English*

Security Management: A Critical Thinking Approach, *Michael Land, Truett Ricks, and Bobby Ricks*

Security Management for Occupational Safety, *Michael Land*

Workplace Safety and Health: Assessing Current Practices and Promoting Change in the Profession, *Thomas D. Schneid*

Forthcoming Titles

Physical Security and Safety: A Field Guide for the Practitioner,
Truett A. Ricks, Bobby E. Ricks, and Jeffrey Dingle

Physical Hazards of the Workplace, Second Edition, *Paul English*

Creative Safety Solutions, Second Edition, *Thomas D. Schneid*

Workplace Safety and Health

Assessing Current Practices and Promoting Change in the Profession

Thomas D. Schneid

CRC Press
Taylor & Francis Group
Boca Raton   London   New York

CRC Press is an imprint of the
Taylor & Francis Group, an **informa** business

CRC Press
Taylor & Francis Group
6000 Broken Sound Parkway NW, Suite 300
Boca Raton, FL 33487-2742

Printed on acid-free paper
Version Date: 20140210

International Standard Book Number-13: 978-1-4398-7410-3 (Paperback)

Library of Congress Cataloging-in-Publication Data

Schneid, Thomas D.
    Workplace safety and health : assessing current practices and promoting change in the profession / author, Thomas D. Schneid.
        pages cm. -- (Occupational safety and health guide series)
    Includes bibliographical references and index.
    ISBN 978-1-4398-7410-3 (paperback)
    1. Industrial safety. I. Title.

T55.S32824 2014
658.3'82--dc23   2014002270

Visit the Taylor & Francis Web site at
http://www.taylorandfrancis.com

and the CRC Press Web site at
http://www.crcpress.com

Contents

Preface

The world around us is changing, but the safety profession often clings to the basic beliefs and practices that evolved in the early years after the promulgation of the Occupational Safety and Health Act of 1970. Are these tried-and-true practices and activities still effective in our changing workplace? Is there a better way of achieving better results in safeguarding our employees from accidents and injuries? In short, why do you perform the safety activities that you do on a daily basis, and are your safety efforts producing the results necessary to keep your safety program and your company competitive in today's global market?

This book examines many of the basic assumptions that we hold "near and dear" in the safety profession, and challenge the reader to assess and evaluate his or her activities in search of a better and more effective way of achieving the results necessary to be competitive in today's workplace. On average, between 8000 and 10,000 workers in the U.S. workplace are killed on an annual basis, and millions of workers suffer debilitating injuries and illnesses. This number of workers injured or killed in the workplace has improved slightly in the past 40 years. However, the average number of workers killed equals the population of a small town, and the number of workers injured equals the population of a large metropolitan city. We need a game changer—a new way of achieving a safe and healthful workplace!

The safety profession is one of the only professions that can have a direct or indirect effect on the lives of workers, their families, communities, and all the many others linked to the worker. This important responsibility is often clouded by the daily safety activities, safety programs, and individual challenges that the safety professional faces on a daily basis. It is the author's hope that this text will permit readers to view their activities and actions from a different prospective and see the real impact that they have on the lives of others and strive to find better ways of creating and maintaining a safe workplace that eliminates all injuries and illnesses.

Acknowledgments

My thanks to the many online and on-campus graduate students in the Master of Science Degree in Safety, Security and Emergency Management courses at Eastern Kentucky University who have challenged the status quo and inspired their classmates and graduate faculty to strive to find a better way of safeguarding the moms and dads who work in the American workplace on a daily basis.

I also thank my parents, Bob and Rosella, for their foresight, sacrifices, and lifelong emphasis on education.

Last, I thank my wife, Jani, and my children, Shelby, Madison, and Kasi, for their time and patience during the writing of this text. Time is definitely a fleeting commodity.

About the Author

Thomas D. Schneid is the chair of the Department of Safety and Security and a tenured professor in the School of Safety, Security and Emergency Management (SSEM) in the College of Justice and Safety at Eastern Kentucky University. In his 25 years at EKU, Tom has served in many capacities, including chair of the Department of Safety and Security, interim chair for SSEM Graduate Studies and Research, graduate program director for the online and on-campus Master of Science Degree in Safety, Security and Emergency Management, coordinator of the Fire and Safety Engineering program, and SSEM Career and Cooperative Education coordinator.

Tom has worked in the safety and human resource fields for over 30 years in various capacities, including corporate safety director and industrial relations director. In Tom's legal practice, he has represented numerous corporations in OSHA and labor-related litigation throughout the United States. Tom earned a BS in education, MS and CAS in safety, as well as his juris doctor (JD) in law from West Virginia University and LLM (graduate law) from the University of San Diego. Tom is a member of the bar of the U.S. Supreme Court, 6th Circuit Court of Appeals, and a number of federal districts, as well as the Kentucky and West Virginia Bars.

Tom has authored numerous texts, including *Corporate Safety Compliance Law, OSHA and Ethics* (2008); *Americans with Disabilities Act: A Compliance Guide* (1994); *ADA: A Manager's Guide* (1993); *Legal Liabilities for Safety and Loss Prevention Professionals* (2010); *Fire and Emergency Law Casebook* (1996); *Creative Safety Solutions* (1998); *Occupational Health Guide to Violence in the Workplace* (1999); *Legal Liabilities in Emergency Management* (2001); and *Fire Law* (1995). Tom has also coauthored several texts, including *Food Safety Law* (1997), *Legal Liabilities for Safety and Loss Prevention Professionals* (1997), *Physical Hazards in the Workplace* (2001), and *Disaster Management and Preparedness* (2000), as well as over 100 articles on safety and legal topics. Tom recently completed work on a new text titled *Labor and Employment Issues for Safety Professionals* and is currently working on a text on legal issues in safety and security.

1 Is There a Loss of Trust in the Workplace?

No virtue is more universally accepted as a test of good character than trustworthiness.

—Harry Emerson Fosdick, D.D.

To be trusted is a greater compliment than to be loved.

—George MacDonald

LEARNING OBJECTIVES

1. Identify the impacts related to trust in the workplace.
2. Identify the issues involving trust in the workplace.
3. Analyze the issue of trust on the safety and health function.
4. Identify issues involving "presenteeism" in the workplace.
5. Analyze the impact of the economy on employee trust in the workplace.

Safety and health professionals will be facing new and challenging issues and situations as a result of the globalization of society, as well as our changing workplace environments. The status quo that has been prevalent in the safety and health profession since the enactment of the Occupational Safety and Health Act in 1970[1] may require substantial revision or modification in order to ensure appropriate application in future workplaces and with the future workforce. Many of the values and virtues in our employees and our companies that safety and health professionals have previously taken for granted may have changed or have been lost as a result of globalization, corporatization, and the economy.

According to the *Merriam-Webster Dictionary*, *Trust* is:

a: assured reliance on the character, ability, strength, or truth of someone or something;
b: one in which confidence is placed.[2]

In the safety profession, trust is a basic commodity that is essential at all levels in order for the safety professional to adequately and appropriately manage the function. Top management rely on the information acquired and provided by the safety professional to guide their decision-making process. Middle or operations level management rely on the expertise and experience of the safety professional to guide their safety and health program level efforts.

The supervisory level relies on the safety professional for education, training, and guidance in their day-to-day safety and health efforts. And above all, each and every employee trusts that the safety professional is providing his or her best efforts in creating and maintaining a safe and healthful workplace. Trust is an essential element at each level within the operational structure in order for the safety and health function to operate at maximum efficiency and effectiveness.

Has the American worker lost trust in his or her company or organization due to the economic realities of downsizing, rightsizing, mergers, takeovers, and bankruptcies?

As the American workplace strains to rapidly adapt in order to address the current economic challenges, globalization, regulatory requirements, aging workforce, and other myriad old and new issues, has trust become a lost element in our corporate "survival mode" mentality, and if trust is lost, how has or will this impact the safety and health function now as well as in the future? Does an organization need trust among and between the various corporate levels and in the safety function to operate efficiently and effectively? Can trust be maintained where there are a multitude of changes and potential inequities in the workplace or regenerated, if lost, as a result of current corporate restructuring? Can the safety and health function continue to rely on this trust at all levels to achieve the expected results, or is there a new dynamic in which trust is not an essential element that safety professionals will need to address in the near future? Can the safety and health function perform effectively where individuals at all levels of the organization adopt a me rather than us approach?

Let me explain my general observations. The American workplace has always been a place of challenges, strife, and confrontation, but also a place of trust, competitiveness, innovation, compassion, and fulfilled promises. In the not so recent past, individuals went to work for a company or organization with the expectation that they would be paid fairly for their efforts, benefits would be provided, their job would be secure if they performed adequately, and at the end of their employment, a secure pension would be provided to assist them financially through their golden years. It was the norm, rather than the exception, that an individual would spend his or her entire career with one company or organization. However, in recent decades, the American workplace and the American society have undergone a change that has dramatically shifted not only the perception, but also the functionality of the average workplace.

In prior decades, companies and organizations planned for the longevity of the company and organizations, and utilized such tools as a 5-year planning cycle to provide a longer view of the direction of the company or organization. As companies and organizations transitioned from privately held companies to publically held companies with Wall Street and shareholder involvement, the management emphasis shifted from the longer view to a quarterly-by-quarter view where immediate profits became priority. This change in corporate America's management focus, along with the changing economic climate, has created a major shift at all levels within the American workplace. The new American workplace within which the safety and health professional will be operating has changed, and the safety professional should be prepared for these changes.

The American workforce is changing. Taking a very broad and generalized view of the entirety of Generation Y, there are many unique characteristics of your current and future employees that may impact the trust area, and thus impact the safety and health profession.

Generation Y has grown up with and utilized current and changing technology, from desktop computers to laptops, and this technology is becoming a necessity rather than a luxury for many of your employees. From the 2010 Pew Research Center report,[3] 93 percent of Generation Y use the Internet, 75 percent have created a profile on a social networking site, 41 percent use only a cell phone (with no landline), and 20 percent have posted a video of themselves online.[4] Your employees from Generation Y have grown up working with these technologies and rely on these technologies on the job and in their personal life. Generation Y is very comfortable with technology and "prefers to communicate through e-mail and text messaging rather than face-to-face contact."[5]

Your Generation Y employees are also the most diverse, highly educated, and willing to challenge the status quo than any previous generation. Safety and health professionals now and in the near future will see a workplace that is more racially and ethnically diverse, employees who are more "knowledgeable, intellectual, and skilled,"[6] and employees who are more mature than in the past. Your Generation Y employees have grown up in homes where both parents have worked and their children needed to grow up quickly with an understanding of their personal priorities. Generation Y employees will only spend the time necessary to successfully complete the job or assignment, but will generally not go the extra mile. Safety and health professionals should be prepared for employees to ask why and fully expect an expansive explanation of the safety and health concepts, theories, and assignments. Safety and health professionals should be aware that the economic environment in which Generation Y has grown up has had a greater impact than on previous generations. Generation Y has seen their parents lose jobs and attempt to acquire a new jobs, lose health care and other benefits, lose pensions, and in many circumstances, lose the very home in which their were raised. Generation Y has been uprooted and moved for the financial well-being of the family. As a result of these experiences, Generation Y has little confidence in corporate America and realizes that corporate America is about profit and not people. "Generation Y has adopted more of a free agent mentality."[7]

Generation Y, with their skills, work ethic, and techno-savvy, will stay at a job only as long as they need it or a better job opportunity presents itself. Generation Y does not trust corporate America and often has no plans to stay with the same employer for an entire career. Generation Y is far more entrepreneurial, willing to risk all to pursue their dreams. Also, safety and health professionals should be aware that the recent economic downturn has also placed Generation Y in debt and often looking for a job. In 2008, "67 percent of students graduating from four-year colleges and universities has student loan debt ... and the average student loan debt level for graduating seniors was $23,000.00."[8] Additionally, 36 percent of Generation Y depend on financial support from their families, and 31 percent, even if working, are uninsured.[9] As a result, Generation Y may have mortgaged their financial future to meet current financial needs.

The average Generation Y student possesses more than three credit cards, and 20 percent carry a balance of over $10,000.00.[10] More than half of all working

Generation Y have cashed out their 401K retirement plans when they changed or lost their jobs.[11]

Safety and health professionals should be aware that Generation Y is far more family oriented, life style centric, and achievement oriented than past generations. However, although achievement oriented, Generation Y employees are team players and value the team experience. As a result of the economic environment, social networking, or even the fact that their parents are providing financial support, Generation Y places a priority on family over work. "The fast track has lost much of its appeal from Generation Y who is willing to trade high pay for fewer billable hours, flexible schedules and a better work/life balance."[12] Safety and health professionals should be cognizant in this stuff in priorities with their workforce.

Lifestyle also takes a priority position in Generation Y employees' lives. As an example, 25 percent of Generation Y are unaffiliated with any religion, and 40 percent have a tattoo.[13] Additionally, only 2 percent of Generation Y are military veterans, and only 21 percent are married (the lowest percentage in three generations).[14] Safety and Health professionals should be cognizant of these types of changes in their workforce and workplace.

Lastly, Generation Y, in addition to being achievement oriented and highly educated and motivated, also crave attention. Instructors should be aware that Generation Y expects, or even demands, guidance and feedback. "It is almost cliché to say that Generation Y is over parented. But they are."[15] Generation Y expects to be "kept in the loop" and frequently praised for their achievements. Generation Y seek their managers reassurance and also look to their instructors for structure, boundaries, and context. The employee often looks to their instructors in substitution for their parents for approval and reassurance. Generation Y students want to be respected but genuinely cared for by management, treated as adults, provided with guidance and benchmarks, and let know that everyone doesn't get a trophy simply for showing up.[16]

Safety and Health professionals should clearly and carefully let Generation Y employees know exactly how they can earn rewards, and the rewards are "tied to concrete actions within their own direct control."[17]

Let's look at this observation with more specificity and ascertain the impact on trust, or lack thereof, in the American workplace. Many companies and organizations have changed from a defined benefit pension program to a 401K (or 403B) retirement system. This type of change saved the company money in future benefits and shifted the responsibility for retirement savings to the individual worker. Although companies and organizations often match employee contributions, contributing and managing the funds became the responsibility of the employee rather than the company or organization. A positive for employees is that the 401K (or 403B) plan is often more transferrable than the traditional pension plan for employees leaving the company or organization. However, with this shift in responsibility for their future retirement to individual employees, is the individual employee knowledgeable and capable of effectively managing these essential future funds? Has the transferability of the 401K (or 403B) created a transient employee with fewer ties to the company or organization? Are companies or organizations adopting a shorter-term outlook with 401K matching systems rather than the longer-term defined pension plan? Have the companies or organizations essentially transferred responsibility for their employee's long-term welfare to the individual employee, and what will be the long-term impact if or when the employee fails to contribute (and acquire the matching funds) to the 401K (or 403B) or mismanages the funds due to lack of knowledge or expertise and becomes destitute after 20 or 30 years of service to the company or organization?

Has or will trust be lost between the company or organization and the employee? Due primarily to economic conditions, many American households over the past decades required both spouses to work in order to support the family unit. Our next generation of employees, their children, watched their parent's jobs get downsized, their parent's companies or organizations bought and sold, their parent's paychecks shrink, their parent's benefits declining, their parents being terminated, and their parents uprooting the family to acquire new jobs. Safety and health professionals should be aware that your employees may be entering the workplace with an inherent mistrust of the corporate environment as a result of their parent's experiences. For safety and health professionals, how will the experiences of Generation Y impact the safety and health function now and in the future?

Historically in the United States, labor organizations or unions often served as a buffer and a level of protection between management and the hourly workforce. To a great extent, there was a trust, security, and anticipated longevity that many employees acquired through this relationship. However, in recent decades, employee representation by labor organizations in the private sector has significantly declined. Absent an individual employment contract or representation under a collective bargaining agreement, most employees were identified by law as "at will" employees, and potentially the balance within the employer-employee relationship significantly shifted toward the employer. Has the trust between the employer and the employee changed as a result of the changing employer-employee relationship? Have the shifting sands of the employer-employee relationship had an impact on the safety and health function in the American workplace?

The American workforce is aging, and older employees want or economically need to remain in the workplace. There is little doubt that the recent economic pressures have forced many employees to rethink the concept of traditional retirement and elect to remain in the workforce. Additionally, given the demographics of a smaller and declining pool of younger workers as well as the increased longevity and physical function of older employees, many companies and organizations may need the experience and expertise of their older employees in the workplace.[*] How will our aging workforce impact the safety and health function within the company or organization? How will laws and regulations, such as the Americans with Disabilities Act (ADA) and Age Discrimination in Employment Act (ADEA), impact the workforce as well as the management thereof?[†]

As a result of the recent economic conditions, many individuals who are employed work in generalized fear of losing their jobs. The issue of presenteeism has created new and challenging issues for safety and health professionals. Presenteeism is when

---

[*] *The Aging Workforce: Challenges and Opportunities for Providers and Employers*, Bruyere and Young, Cornell University, Ithaca, NY (2012). Also see www.ilr.cornell.edu.
[†] Id.

employees "show up to work even when they should be home.... Individuals are ill, potentially contagious, and not functioning at 100 percent, but they still feel they should be in the office.... Presenteeism can also apply to people who work late or come into the office during their vacations."* In addition to a generalized fear of losing one's job in tough economic times, other factors, such as the lack of sick days, workplace culture, and workplace policies, can also result in contagious employees contaminating the workplace.† How can presenteeism impact your safety and health programs and efforts? Can a contagious employee cause more harm to the company or organization by coming to work than not?

With the multitude of changes that have and will take place within the health care insurance arena as a result of recent legislation, current and future employees may be uncertain as to the impact these changes may have on their individual health care situation. Employees currently covered under an employer-based health care plan have seen the costs of their coverage substantially increase and the benefit level decline. Potential employees may base their decision whether or not to work for your company or organization on the health care coverage or lack thereof. How will this uncertainty impact your safety and health programs? Will the changes in the health care laws impact your management of workers' compensation?

Given the observations above as well as myriad additional employment-related matters surrounding this issue, the fundamental question lying at the heart of the possible workplace transition is whether the historical trust that was created between the company and the employee continues to exist, and what impacts, if any, will this loss of trust have on the safety and health function. Has the American workplace become a free agent market wherein the employer views employees as a commodity with profit taking priority over people? Do employees view the job as a short-term layover while waiting for better opportunities? Has the trust bond between the employer and employee that was instrumental in the success of American industry for many decades been permanently severed as a result of globalization of jobs, the current economic recession, and related factors, or is this simply an anomaly in the labor-management relationship cycle?

If arguably the American workplace has or is undergoing a change and the historical trust relationship between the employer and the employee has or is being altered, the safety and health function in many companies and organizations may be at the center of metamorphosis. As we are aware, the safety and health function does not operate in a vacuum and not interact with many of the other functions within the organization or company. Can the safety and health function successfully function within an environment where employees possess an inherent mistrust of the corporate or organizational entity? In a traditional environment where productivity always takes priority, can the safety and health function achieve the requisite environment necessary to properly safeguard all employees in the workplace? Will the safety and health professional need to adapt and change to meet the needs of the current environment in order to achieve success?

---

* "The Cost of Presenteeism," Patricia Puckett, at www.about.com; also see "The Hidden Cost of Presenteeism: Causes and Solutions," Patricia Schaefer, at www.businessknowhow.com.
† Id.

If the hypothesis is that the trust relationship between the company or organization and the employee has changed or has been lost, the question for safety and health professionals is: How must the safety and health function also be changed or modified in order to address this inherent cultural shift in the workplace? Trust between the company or organization and its employees often takes years to build and can be lost in a minute through actions or inactions by either party in the relationship. How will you, as the safety and health professional, address these changes and maintain a safe and healthful environment for all employees?

DISCUSSION QUESTIONS

1. What is trust and how can it impact the safety and health function?
2. How has the economy impacted the American workplace? The safety and health function?
3. What is presentteeism and how does it impact the safety and health function?
4. How is the American workplace changing and what is the impact on safety and health function?

ENDNOTES

1. Pub. L. 91-596, 84 STAT. 1590, 91st Congress, S.2193, December 29, 1970, as amended through January 1, 2004.
2. *Merrian-Webster Online Dictionary* at www.Merriam-Webster.com.
3. Pew Research Center 2010 report, *Millennials: A Portrait of Generation Next*; also see "36 Facts about Generation Y in the Workplace and Beyond," Rosetta Thurman, at www.rosettathurman.com2010/07/36; "Number Crunching: The Top 51 Stats for Generation Y Marketers," Ekaterina Walter, at www.thenextweb.com on January 21, 2012.
4. Ibid.
5. "Generation Y," Sally Kane, at About.com.
6. "Gen Y Characteristics: Stereotype of Generation Y," Life and Luxury, at www.lifeandluxury.hubpages.com.
7. Ibid.
8. "36 Facts about Generation Y in the Workplace and Beyond," Thurman.
9. Ibid.
10. Ibid.
11. Ibid.
12. "Generation Y," Kane.
13. "36 Facts about Generation Y in the Workplace and Beyond," Thurman.
14. Ibid.
15. "Not Everyone Gets a Trophy—How to Manage Generation Y," Bruce Tulgan, *Loss Prevention Magazine*, September–October 2012, pp. 51–58.
16. Ibid.
17. Ibid.

2 Why Compliance?

While all other sciences have advanced, that of government is at a standstill—little better understood, little better practiced now than three or four thousand years ago.

—John Adams

Governments, like clocks, go from the motion men give them, and as governments are made and moved by men, so by them they ruined also. Therefore governments depend upon men rather than men upon governments.

—William Penn

LEARNING OBJECTIVES

1. Identify and analyze the structure developed under the OSH Act.
2. Analyze the methodology in developing standards.
3. Assess the important of compliance within the safety and health function.
4. Analyze and assess I2P2.

Since the enactment of the Occupational Safety and Health (OSH) Act in 1970, the safety and health function has been primarily based on compliance and the myriad safety and health standards promulgated by the Occupational Safety and Health Administration (OSHA). The standards developed during the early years of the OSH Act were primarily developed through review of other voluntary standards and recommendations by other organizations that were already in place. New standards were developed over the years through research and as a reaction to injuries and illnesses that had been incurred in the workplace. However, with numerous safety and health standards in place and being strictly enforced, the American workplace still experiences approximately 4000 fatalities on an annual basis,[*] and workplace injuries and illnesses cost employers many millions of dollars each year.

Since its inception four decades ago, the Occupational Safety and Health Administration and state plan programs have had a dramatic impact on safety and health in the American workplace. "Workplace fatalities have been reduced by more than 65 percent and occupational injury and illness rates have declined by 67 percent,"[†] while employment in the American workplace has doubled. On-the-job fatalities have been substantially reduced from "38 worker deaths per day in 1972 to

---

[*] 4609 fatalities in 2001. See OSHA website: www.osha.gov.

[†] OSHA website: www.osha.gov.

13 per day in 2011."[*] Injury and illness rates have been reduced from "10/9 incidents per 100 workers in 1972 to fewer than 4 per 100 workers in 2010."[†]

Through OSHA's development of broad "general industry" and construction standards that encompass virtually all industries as well as more specific standards addressing specific industries or hazards, a foundational base for compliance has been established. This expansive volume of standards and regulations addressing hazards in the workplace is strictly enforced through site inspections[‡] and penalties for noncompliance that can include monetary as well as criminal penalties.

However, the question for many safety and health professionals today is whether simply complying with the OSHA standards is sufficient to safeguard all employees in the workplace? Compliance with the OSHA standards may address most of the major hazards in many workplaces (and is required); however, is the achievement of a sufficient compliance level with the appropriate OSHA standards sufficient to effectively and efficiently safeguard the workplace and reduce the number and severity of potential occupational injuries and illnesses (and the correlating workers compensation cost)? Are safety and health professionals being employed by companies and organizations to achieve compliance (and avoid potential OSHA penalties), or is there an expectation of a minimization or elimination of all hazards and the correlating reduction in losses and costs?

In an effort to address specific risks, reduce correlating costs, and be able to appropriately manage the safety and health function, many safety and health professionals developed extensive written compliance programs through which compliance is achieved and maintained as part of the all-encompassing safety and health effort that also address the specific risks and hazards inherent in their unique workplaces. Utilizing the OSHA standards as a foundation, many safety and health professionals developed and implemented numerous additional programs to address the specific needs of their workplace and to escalate the level of protection for their employees.

Safety and health professionals should be aware that the standards promulgated by OSHA provide the "bare bones" minimum level of safety and health under which an employer may not fall, and the standards are written in a manner to encompass all employers within OSHA's jurisdiction. Standards have been added over the past 40 years to address identified risks and hazards, with each standard basically "standing alone" with specific standard requirements. Safety and health professionals often developed specific written programs for each specific hazard and standard applicable to their workplace for the purposes of compliance with the applicable standard. However, often, to achieve company goals in the areas of injury and illness reduction, emerging or unique hazards, cost reductions, or other priorities, safety and health professionals often incorporate the stand-alone compliance programs into a broader, more comprehensive, and systematic managerial approach through which to manage the entire safety and health function.

OSHA has developed a number of different programs over the years, including, but not limited to, the Voluntary Protection Program (VPP), Safety and Health

---

[*] Id.

[†] Id.

[‡] Id. The total federal inspections in 2011 was 40,648, and state plan inspections totaled 52,056.

Achievement Recognition Program (SHARP), and OSHA Challenge Program,[*] which recognized employers for their voluntary safety and health efforts. However, there were no specific standards that required an employer to develop injury and illness programs beyond the individual standards developed to address specifically identified hazards. With the employer virtually always responsible for violations under the OSH Act and employers usually responsible for the cost of on-the-job injuries and illnesses under individual state workers' compensation laws, the safety and health emphasis by many companies and organizations shifted over the years from compliance and fear of OSHA to cost reduction through injury and illness prevention.

OSHA has recently proposed the adoption of the Injury and Illness Prevention Program (I2P2) to provide employers with a proven method through which to reduce workplace injuries and illnesses. In general, the proposed Injury and Illness Prevention Program is a "flexible, commonsense, proven tool to find and fix hazards *before* injuries, illnesses, or death occurs."[†] The Injury and Illness Prevention Program incorporates six core elements through which to manage the safety and health function, including management leadership, worker participation, hazard identification and assessment, hazard prevention and control, education and training, and program evaluation and improvement.[‡] Included in this chapter is the Injury and Illness Prevention Program's White Paper,[§] which provides a summary of the proposed requirements, as well as the cost of injuries and illnesses and examples of successful injury and illness prevention efforts.

Although many safety and health professionals may already be addressing injury and illness prevention as part of their overall compliance efforts or comprehensive safety and health program, prudent safety and health professionals may wish to incorporate these core elements into existing programs or incorporate an injury and illness prevention program into your existing safety and health program function. Although compliance with the OSHA standards is often the foundation of our safety and health efforts, ensuring compliance with the standards is often not enough to safeguard employees in the workplace. Additionally, employers today are very cognizant of the increasing costs of on-the-job injuries and illnesses, and the safety and health function is being driven by the volume and costs of workplace injuries and illnesses rather than the fear of penalties by OSHA. Compliance with the OSHA standards is the foundation, mandatory, and does not going away any time soon. However, today's safety and health profession should strive to achieve prevention levels that far exceed the foundational levels of compliance to truly safeguard employees in the workplace as well as achieve the correlating cost savings for their employers. The current levels of +4000 workplace deaths and millions of workplace injuries and illnesses annually are simply not acceptable by OSHA, employers, and employees and should not be by safety and health professionals. Your daily efforts, as a safety and health professional, do have an impact on the lives of your employees, your community, your company, and many more.[¶]

---

[*] OSHA website: www.osha.gov.
[†] Id.
[‡] Id.
[§] Id. White Paper published in January 2012.
[¶] OSHA website: www.osha.gov.

Injury and Illness Prevention Program "White Paper" from OSHA website.

Injury and Illness
Prevention Programs
White Paper
January 2012

Introduction/Executive Summary

An *injury and illness prevention program*,1 is a proactive process to help employers find and fix workplace hazards before workers are hurt. We know these programs can be effective at reducing injuries, illnesses, and fatalities. Many workplaces have already adopted such approaches, for example as part of OSHA's cooperative programs. Not only do these employers experience dramatic decreases in workplace injuries, but they often report a transformed workplace culture that can lead to higher productivity and quality, reduced turnover, reduced costs, and greater employee satisfaction.

Thirty-four states and many nations around the world already require or encourage employers to implement such programs. The key elements common to all of these programs are management leadership, worker participation, hazard identification and assessment, hazard prevention and control, education and training, and program evaluation and improvement.

Based on the positive experience of employers with existing programs, OSHA believes that injury and illness prevention programs provide the foundation for breakthrough changes in the way employers identify and control hazards, leading to a significantly improved workplace health and safety environment. Adoption of an injury and illness prevention program will result in workers suffering fewer injuries, illnesses and fatalities. In addition, employers will improve their compliance with existing regulations, and will experience many of the financial benefits of a safer and healthier workplace cited in published studies and reports by individual companies, including significant reductions in workers' compensation premiums.

Background

In the four decades since the *Occupational Safety and Health Act* (OSH Act) was signed into law, workplace deaths and reported occupational injuries have dropped by more than 60 percent. Yet the nation's workers continue to face an unacceptable number of work-related deaths, injuries and illnesses, most of them preventable:

- Every day, more than 12 workers die on the job – over 4,500 a year.
- Every year, more than 4.1 million workers suffer a serious job-related injury or illness.

An enhanced focus on prevention is needed to bring these numbers down. To accomplish this, an effective, flexible, commonsense tool is available that can dramatically reduce the number and severity of workplace injuries and illnesses: the injury and illness prevention program. This tool helps employers find hazards and fix them *before* injuries, illnesses or deaths occur. It helps employers meet their obligation under the OSH Act to "furnish to each of his employees employment and a place of employment which are free from recognized hazards that are causing or are likely to cause death or serious physical harm to his employees." It also helps employers avoid the significant costs associated with injuries and illnesses in the workplace.

Injury and illness prevention programs are not new, nor are they untested. Most large companies whose safety and health achievements have been recognized through government or industry awards cite their use of injury and illness prevention programs as their key to success. Convinced of the value, effectiveness, and feasibility of these programs, many countries around the world now require employers to implement and maintain them. These countries include Canada, Australia, all 27 European Union member states, Norway, Hong Kong, Japan and Korea. This initiative also follows the lead of 15 U.S. states that have already implemented regulations requiring such programs.

How Does an Injury and Illness Prevention Program Work?

Most successful injury and illness prevention programs include a similar set of commonsense elements that focus on finding all hazards in the workplace and developing a plan for preventing and controlling those hazards. Management leadership and active worker participation are essential to ensuring that all hazards are identified and addressed. Finally, workers need to be trained about how the program works and the program needs to be periodically evaluated to determine whether improvements need to be made.

These basic elements – management leadership, worker participation, hazard identification and assessment, hazard prevention and control, education and training, and program evaluation and improvement – are common to almost all existing health and safety management programs. Each element is important in ensuring the success of the overall program, and the elements are interrelated and interdependent.

When it comes to injury and illness prevention programs, every business is different, and one size certainly does not fit all. Employers who implement injury and illness prevention programs scale and adapt these elements to meet the needs of their organizations, depending on size, industry sector or complexity of operations.

What Are the Costs of Workplace Injuries, Illnesses and Deaths to Employers, Workers and the Nation?

The main goal of injury and illness prevention programs is to prevent workplace injuries, illnesses and deaths, the suffering these events cause workers, and the financial hardship they cause both workers and employers.

Workplace incidents cause an enormous amount of physical, financial and emotional hardship for individual workers and their families. Combined with insufficient workers' compensation benefits and inadequate medical insurance, workplace injuries and illnesses can not only cause physical pain and suffering but also loss of employment and wages, burdensome debt, inability to maintain a previous standard of living, loss of home ownership and even bankruptcy. When implemented effectively, injury and illness prevention programs can help workers and their families avoid these disruptive and sometimes calamitous impacts on their lives.

At the same time, these programs will help employers avoid the substantial cost impacts and business disruptions that accompany occupational injuries, illnesses and deaths. One widely-cited source regarding estimates of the magnitude of these costs is the Liberty Mutual Research Institute, which reports the direct cost of the most disabling

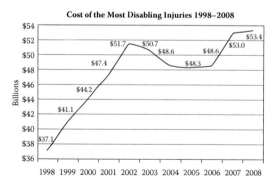

Cost of the Most Disabling Injuries 1998–2008

Source: Liberty Mutual Research Institute, 2010.

workplace injuries in 2008 to be $53 billion (Liberty Mutual Research Institute, 2010).2 Another source, the National Academy of Social Insurance (NASI), estimates **the annual workers' compensation benefits paid for all compensable injuries and** illnesses in 2009 at $58 billion (National Academy of Social Insurance, 2011). NASI further reports the total costs paid by employers for workers' compensation increased from $60 billion in 2000 to $74 billion in 2009.

In addition to these direct costs, employers incur a variety of other costs that may be hidden or less obvious when an employee is injured or ill, but in most cases involve real **expenditures of budget or time. These expenditures are commonly referred to as** *indirect* costs and can include:

- Any wages paid to injured workers for absences not covered by workers' compensation;
- **The wage costs related to time lost through work stoppage;**
- Administrative time spent by supervisors following injuries;
- Employee training and replacement costs;
- Lost productivity related to new employee learning curves and accommodation of injured employees; and
- Replacement costs of damaged material, machinery and property.

OSHA has historically used the results of one study (Stanford University, 1981) that found the indirect costs can range from 1.1 (for the most severe injuries) to 4.5 (for the least severe injuries) times the direct costs.3

When workers are killed, are injured or become ill, there are substantial costs beyond those borne by employers. A variety of approaches can be used to estimate these costs. For example, Viscusi and Aldy (2003) provided estimates of the monetary value of each life lost. OSHA

> "Establishing safety as a value rather than a priority tells our employees and our customers that safety is built into our culture, not something we do to merely comply with regulations.
>
> Our excellent safety performance over the past seven years has been a key factor in reducing our insurance cost. Our low EMR [Experience Modification Rate], incidents rates, and SHARP Management System have impressed our customers and, in many cases, was a key factor in selecting Parsons to perform their project."
>
> – *Charles L. Harrington, Chairman* & *CEO, Parsons Corp.*
>
> **Source: National Safety Council.**

updated these estimates (to account for inflation) to 2010 dollars, yielding a value of $8.7 million for each life lost. Multiplying this value by the 4,547 workplace deaths reported by the Bureau of Labor Statistics for 2010, OSHA estimates the annual cost of known workplace fatalities to be nearly $40 billion.

This estimate does not include the cost of non-fatal injuries, or of occupational illnesses like cancer and lung disease. These illnesses generally may occur many years or even decades after workers are exposed and are therefore seldom recorded in government statistics or employer surveillance activities.

The human and economic costs of these conditions are indisputably enormous. Leigh et al. (1997) estimated that more than 60,000 workers die each year from occupational illnesses, and more than 850,000 develop new illnesses annually. Similarly, Steenland et al. (2003) estimated that between 10,000 and 20,000 workers die each year from cancer due to occupational exposures, and between 5,000 and 24,000 die from work-related Chronic Obstructive Pulmonary Disease.

In summary, the number and costs of workplace injuries, illnesses and fatalities are unacceptably high. Injury and illness prevention programs have been proven to help employers and society reduce the personal, financial and societal costs that injuries, illnesses and fatalities impose. As described below, the thousands of workplaces that have implemented these programs in some form have already witnessed the resulting benefits, in the form of higher efficiency, greater worker productivity and lower costs.

What Is the Evidence that Injury and Illness Prevention Programs Protect Workers and Improve the "Bottom Line"?

Numerous studies have examined the effectiveness of injury and illness prevention programs at both the establishment and corporate levels (e.g., Alsop and LeCouteur, 1999; Bunn et al., 2001. Conference Board, 2003; Huang et al., 2009; Lewchuk, Robb, and Walters, 1996; Smitha et al., 2001; Torp et al., 2000; Yassi, 1998). This research demonstrates that such programs are effective in transforming workplace culture; leading to reductions in injuries, illnesses and

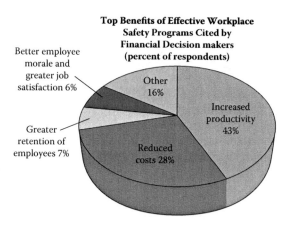

Top Benefits of Effective Workplace Safety Programs Cited by Financial Decision makers (percent of respondents)

fatalities; lowering workers' compensation and other costs; improving morale and communication; enhancing image and reputation; and improving processes, products and services. The studies also highlight important characteristics of effective programs, including management commitment and leadership, effective employee participation, integration of health and safety with business planning and continuous program evaluation. They suggest that programs without these features are not as effective (Shannon et al., 1996, 1997; Gallagher, 2001; Gallagher et al., 2003; Liu et al., 2008).

One study (Smitha et al., 2001) focused on manufacturing facilities in 13 states with mandatory injury and illness prevention programs and/or mandatory health and safety committee requirements. The authors found that both types of regulations were effective in reducing injury and illness incidence rates. Three of the four states with only safety and health program requirements experienced the greatest reductions in injury and illness rates following promulgation of these mandatory program regulations.

OSHA examined the injury and illness prevention programs in eight states where the state had either required a program or provided incentives or requirements through its workers' compensation programs. The successes of these state programs, which lowered injury and illness incidences by 9 percent to more than 60 percent, are discussed below: Source: Huang et al., 2009. Data based on responses from 231 U.S. companies with 100 or more employees.

- **Alaska** had an injury and illness plan requirement for over 20 years (1973 to 1995). Five years after the program was implemented, the net decrease in injuries and illnesses (i.e., the statewide reduction in injuries and illnesses over and above the national decrease during the same time period) for Alaska was 17.4 percent.
- **California** began to require an injury and illness prevention program in 1991. Five years after this requirement began, California had a net decrease in injuries and illnesses of 19 percent.
- **Colorado** has a program that allows firms to adopt basic injury and illness prevention program components in return for a workers' compensation premium reduction. The cumulative annual reduction in accidents was 23 percent and the cumulative reduction in accident costs was between 58 and 62 percent.
- **Hawaii** began to require employers to have injury and illness prevention programs in 1985. The net reduction in injuries and illnesses was 20.7 percent.
- **Massachusetts** Workers' Compensation program firms receive a premium credit for enrolling in a loss management program. In the first year of this program, firms participating in the program had a 20.8 percent improvement in their loss ratios.
- **North Dakota** has a program under its workers' compensation program for employers who have a risk management program. The incentive is a 5 percent discount on annual workers' compensation premiums. These risk management programs contain many of the elements of an injury and illness prevention program. They resulted in a cumulative decline for serious injuries of 38 percent over a four-year period.
- **Texas** had a program under its workers' compensation commission from 1991 to 2005 which identified the most hazardous workplaces. Those employers were required to develop and implement injury and illness prevention programs. The reduction in injuries, over a four-year period (1992-1995), averaged 63 percent each year.
- **Washington** began requiring establishments to have injury and illness prevention programs in 1973. Five years later the net decrease in injuries and illnesses was 9.4 percent.

OSHA also examined fatality rates and found that California, Hawaii and Washington, with their mandatory injury and illness prevention program requirements, had workplace fatality rates as much as 31 percent below the national average in 2009.

Liu et al. (2008) examined the effectiveness of Pennsylvania's voluntary program that provides workers' compensation premium discounts to employers that establish joint labor-management safety committees. These committees are responsible for implementing several injury and illness prevention program elements: hazard identification, workplace inspection and safety management. The authors found that among program participants there was a strong association between improved injury and illness experience and the level of compliance with the program requirements. This is further evidence that programs with strong management commitment and active worker participation are effective in reducing injury risk, while "paper" programs are, not surprisingly, ineffective.

The literature on injury and illness prevention programs also includes numerous studies that attempt to identify the critical success features associated with superior health

> "There are many benefits from developing a safety culture at your company - none of which is more valuable than employee loyalty. When employees know you care about their personal well-being and you prove that to them in their workplace, it increases morale, engagement, awareness, motivation and productivity."
>
> – *Daniel R. Nobbe, Plant Leader, Fiberteq LLC, Danville, IL.*
>
> **Source: National Safety Council.**

and safety performance. Gallagher (2001) concludes that management commitment and employee involvement are the keys to program success: "[R]ecurring findings across these studies were the critical role played by senior managers in successful health and safety management systems, and the importance of effective communication, employee involvement and consultation."

Worker participation, a fundamental element of injury and illness prevention programs, makes an important contribution to an employer's bottom line. When workers are encouraged to offer their ideas and they see their contributions being taken seriously, they tend to be more satisfied and more productive (Huang et al., 2006). Engaging employees in dialogue with management and each other about safety and health can lead to improved relationships and better overall communication, along with reduced injury rates. Improved employee morale and satisfaction translates to greater loyalty, lower absenteeism and higher productivity.

This body of research, combined with studies of individual companies (see boxes, below, with Case Studies of Programs Implemented under OSHA's Voluntary Protection Program (VPP) and Safety and Health Achievement Recognition Program (SHARP)) demonstrate clearly that injury and illness prevention programs are effective at the establishment level in dramatically reducing risk of workplace injury. This effect has also been detected in state-wide comparisons.

Based on its review of the literature on the effectiveness of these programs and on the experience of the states that have implemented injury and illness prevention program requirements, OSHA estimates that implementation of injury and illness prevention programs will reduce injuries by 15 percent to 35 percent for employers who do not now have safety and health programs. At the 15 percent program effectiveness level, this saves $9 billion per year in workers' compensation costs; at the 35 percent effectiveness level the savings are $23 billion per year.4 In addition to these workers' compensation savings, employers could also save indirect costs incurred when an employee is injured or ill. Beyond the monetized benefits of injuries and illnesses averted, and lives saved, nonmonetized costs of workplace injuries and deaths include uncompensated lost wages, the loss of human capital assets, the loss of productivity, the cost of other government benefits required by injured workers or their survivors, the loss of government tax revenues, other business expenses, and other losses not compensated by workers' compensation or other insurance.

How Widespread are Injury and Illness Prevention Programs?

Employers across the United States have implemented injury and illness prevention programs, and many jurisdictions, in the United States and abroad, currently require or encourage implementation of these programs. Currently, 34 U.S. states have established laws or regulations designed to require or encourage injury and illness prevention programs, including 15 states with mandatory regulations for all or some

Photo: Elena Finizio, Braintree, MA Area Office

employers.5 Other states, while not requiring programs, have created financial incentives for employers to implement injury and illness prevention programs. In some instances this involves providing – or facilitating – workers' compensation insurance premium reductions for employers who establish programs meeting specified requirements. And 16 states, in all three of these groups, provide an array of voluntary guidance, consultation and training programs, and other assistance aimed at helping and encouraging employers to implement injury and illness prevention programs. Depending on the state, these programs apply to all employers, employers above or below a certain size threshold, employers with injury and illness rates above industry average, employers in "high-hazard" industries or employers with above-average workers' compensation experience modification rates.

Summary of Existing State Programs

State	Mandatory Regulation	Mandatory Safety Committees	Consulting or Recognition	Insurance Premium Reductions	If mandatory, who is covered?[a]
Alabama		▓	▓		All employers
Arkansas					"Hazardous" employers
California	▓				All employers
Colorado					
Connecticut		▓			Employers with >25 employees "Hazardous" small employers
Delaware	▓			▓	
Hawaii					All employers
Idaho			▓		
Indiana			▓		
Kansas			▓		
Louisiana	▓			▓	Employers with >15 employees

State	Requirement
Michigan	Employers in construction industry
Minnesota	Employers with >25 employees Committees required for "Hazardous" employers
Missouri	All employers
Mississippi	
Montana	Employers with >5 employees
North Carolina	"Hazardous" employers Committees required for employers with >5 employees
North Dakota	
Nebraska	All employers
New Hampshire	Employers with >10 employees Committees required for employers with >5 employees
New Mexico	
Nevada	Employers with >10 employees Committees required for employers with >25

State	Description
	employees
New York	Employers with payroll >$800K Other "hazardous" employers
Ohio	
Oklahoma	
Oregon	All construction employers All other employers with >10 employees (except logging and agriculture)
Pennsylvania	
Tennessee	"Hazardous" employers
Texas	
Utah	"Hazardous" employers
Vermont	"Hazardous" employers
Washington	All employers
West Virginia	"Hazardous" employers
Wyoming	

a States define "hazardous" employers individually, using criteria such as above-average injury incidence rates for their industry or above-average workers' compensation claim experience.
Source: OSHA Directorate of Standards and Guidance.

The more than 2,400 establishments that belong to OSHA's Voluntary Protection Program have programs that are based on the same core elements found in the injury and illness prevention program that OSHA will be proposing. The same is true for OSHA's Safety and Health Achievement Recognition Program, in which more than 1,500 smaller employers are enrolled. Each year, dozens of organizations seeking international recognition for their safety and health program proudly submit applications to the National Safety Council for the Robert W. Campbell award (see text box). Case studies of past winners are available on the Campbell Award website.

Recognizing Business Excellence in Safety and Health

The Robert W. Campbell Award recognizes organizations that achieve business excellence by integrating environmental, health and safety (EHS) management into **their business operating systems. The Award aims to:**

· Recognize businesses that uphold EHS as a key business value and link **measurable achievement in EHS performance to productivity and profitability.**
· Establish a validated process by which industries can measure the performance of their EHS operations system against well-tested and internationally accepted key performance indicators.
· Use a rigorous systematic review process to capture and evaluate the successes and lessons learned.
· Share leading edge EHS management systems and best practices for educational purposes worldwide.

The Award program is supported by a network of 22 Global Partners across five continents committed to promoting EHS as an integral component of business management worldwide.

Source: www.campbellaward.org.

There are at least two industry consensus standards for injury and illness prevention programs. The American National Standards Institute (ANSI) and American Industrial Hygiene Association (AIHA) have published a voluntary consensus standard, ANSI/AIHA Z10 – 2005 *Occupational Safety and Health Management Systems* (ANSI/AIHA, 2005). The Occupational Health and Safety Assessment Series (OHSAS) Project Group, a consortium of selected Registrars, national standards bodies, professional associations and research institutes, has produced a similar document, OHSAS 18001 – 2007 Occupational Health and Safety Management Systems (OHSAS Project Group, 2007). These consensus-based standards have been widely accepted in the world of commerce and adopted by many businesses on a voluntary basis.

Canada, Australia and all members of the European Union operate programs that either require employers to adopt injury and illness prevention programs, or provide incentives or recognition to those who do so. For example, under the 1989 EU Framework Directive (89/391), EU member countries must have national legislation in place requiring employers to maintain risk identification and prevention programs that are very similar to OSHA's injury and illness prevention program concept (European Union, 1989). U.S. companies operating internationally are familiar with these requirements and have already put in place their own programs to meet these requirements. Finally, many private workers' compensation carriers offer incentives to employers who have injury and illness prevention programs and provide technical assistance to help them implement their programs.

The United States Departments of Defense (DOD) and Energy (DOE) have both adopted this approach for protecting workers employed or stationed at the nation's military installations and nuclear weapons factories, including DOE's high hazard establishments. The success of DOD's program is described in the box below. DOE's program, entitled Integrated Safety

Photo: Frank Wenzel, Washington DOSH

Management, includes an expectation that the facilities will "embrace a strong safety culture where safe performance of work and involvement of workers in all aspects of work performance are core values that are deeply, strongly, and consistently held by managers and workers." According to DOE, the aspects of this safety culture that impact safety performance are Leadership, Employee/ Worker Involvement and Organizational Learning (DOE, 2011).

Despite the value to employers and workers in terms of injuries prevented and dollars saved, many U.S. workplaces have not yet adopted injury and illness prevention programs. Based on the positive experience of employers with existing programs, OSHA believes that injury and illness prevention programs provide the foundation for breakthrough changes in the way employers identify and control hazards, leading to significantly improved workplace health and safety environments. Adoption of injury and illness prevention program will result in workers suffering fewer injuries, illnesses and fatalities. In addition, employers will improve their compliance with existing regulations, and will experience many of the financial benefits of a safer and healthier workplace described in the literature and in reports by individual companies.

The Department of Defense Embraces Injury and illness Prevention Programs

DOD is committed to keeping workers safe from preventable injuries, and has embraced the safety and health management system approach through its participation in OSHA's Voluntary Protection Programs (VPP). The leaders of our armed forces understand that employees are critical to mission readiness, and recognize the link between lost time injuries and illnesses and lost productivity. The Secretary of Defense has set a goal of reducing preventable injuries by 75 percent from a 2002 baseline, with the ultimate aim of achieving zero injuries. VPP participation has proven a powerful tool in this effort.* The 2009 DOD Safety Perception Survey of Senior Leaders captured many positive comments on VPP Successes. According to the head of the Defense Safety Oversight Council (DSOC), which manages DOD's VPP Program, DOD saw a lost day rate reduction of 41 percent, from 31.5 per 100 full-time workers in FY 2002 (before any VPP programs were implemented) to 18.7 per 100 workers in FY 2009. DSOC publishes a list of the "Top 40" installations with the highest lost day rates. One installation that ranked among the highest of these dropped to one of the lowest in under two years through implementation of VPP. The chart below illustrates some of the dramatic improvements in service-wide injury and illness rate performance, comparing data from before and after VPP participation.

VPP Implementation Impacts on Service-Wide Lost Day Rates (per 100 workers)

	FY 02	FY 09	Rate Reduction	Percent Improvement
All DOD	31.5	18.7	12.8	41%
Army	29.3	17.8	11.5	39%
Navy	39.8	21.2	18.6	46%
Marines	73.8	36.7	37.1	50%
Air Force	25.6	16.5	9.1	36%
Defense Logistics	25.6	16.9	8.7	34%

Source: Angello, 2010.

* As of November 30, 2011 there were 39 DOD sites in VPP and approximately 200 additional sites working towards VPP status (Source: OSHA Directorate of Cooperative and State Programs, 2011).

**Case Studies of Programs Implemented under OSHA's
Voluntary Protection Programs (VPP)**

- Hypotherm is a 900-employee, New Hampshire-based manufacturer of high-tech plasma and laser-cutting tools and machines. The company provides an extensive employee training program that emphasizes health and safety as part of an overall focus on quality. Through this investment the firm's highly skilled, safety-oriented workforce has driven a 25 percent reduction in costly machine crashes and down time, and over a 3-year period (2007-2010), the company's workers' compensation costs have fallen by 90 percent. Hypotherm has consistently been named a "Best Place to Work" in the state of New Hampshire and plans to add 100 positions over the next year.
- Allegheny Energy's LM6000 Group operates three combustion turbine facilities in southwestern Pennsylvania. Facing complaints about the use of arc flash hoods required for certain operations (fogging, visibility), the company asked a group of employees to investigate alternatives. The employees identified, evaluated and recommended a power ventilated hood, which the company then purchased. In another case, employees were provided time and resources to identify a way to incorporate fall protection in one particular area. The employees found several locations where vertical lifeline systems could be safely installed and used, and a vendor was brought in to assist with the installation. Involving employees and giving them a role in finding solutions has helped Allegheny Energy foster a culture of safety and remain incident-free since the group began operation.
- Pittsburgh-based McConway & Torley has been producing steel castings, rail couplings, and car-connecting systems for the railroad industry since 1868. The company believes it has the best foundry workers in the world, but also realized that its compliance-focused approach to safety was not enough to prevent workers from getting injured. Working with OSHA, the company began filling gaps in its injury and illness prevention program by following the VPP model. During the process of implementing the VPP program at its two foundries, managers and workers discovered that the required high level of employee involvement really made a difference. With top management's full commitment and support, foundry managers and employees work together to proactively resolve safety issues like repetitive motion problems, to improve work practices and to develop job safety analyses. Employees participate in monthly safety audits, facility-wide inspections, accident investigations and self assessments, and are actively involved in conducting safety training. They feel free to submit ideas for safety improvements – and then they help implement those improvements, a degree of empowerment that continues to make a difference in injury reduction and a safer workplace. The impact of the VPP program was powerful: between 2006 and 2010, McConway & Torley was able to reduce workers' compensation cases in its facilities by 79 percent and reduce related direct costs by 90 percent.

Source: OSHA Directorate of Cooperative and State Programs.

Are Injury and Illness Prevention Programs Too Complicated and Expensive for Small Businesses?

For many small businesses, establishing an injury and illness prevention program may seem daunting. Any program based on formal structures can be difficult to establish in a small organization because of tight budgets. Yet simple, low-cost approaches have been shown to be effective in small businesses (Hasle and Limborg, 2006). Injury and illness prevention programs lend themselves to such low-cost approaches because they are highly flexible – the core elements can be implemented at a basic level suitable for the smallest business, as well as at a more advanced, structured level that may be needed in a larger, more complex organization.

OSHA's Safety and Health Achievement Recognition Program (SHARP), which recognizes small employers that operate exemplary injury and illness prevention programs, provides compelling evidence that such programs can and do work for small businesses. For example, the Ohio Bureau of Workers' Compensation (2011) analyzed the policies of 16 SHARP employers over a 12-year period from 1999 to 2010. The study compared the employers' experience prior to and after achieving entry into the SHARP program. The preliminary results of the study show that the average number of claims for these employers decreased by 52 percent, the average claim cost decreased by 80 percent, the average lost time per claim decreased by 87 percent, and claims (per million dollars of payroll) decreased by 88 percent.

An internal OSHA study of nine SHARP firms, ranging in size from 15 to 160 employees, found that the firms achieved the following as a result of their programs:

- A reduction in the number of injuries and illnesses.
- Improved compliance with regulatory requirements.
- Improved business and cost savings including reduced workers' compensation premiums, reduced administrative and human resources burden associated with filing injury and illness reports, managing workers' compensation cases and training new employees. The companies also experienced improved efficiency in operations and material use, and improved productivity. They were able to leverage their limited health and safety resources.
- An improved workplace environment with greater collective responsibility for workplace health and safety.
- Improved reputation and image in the community including relationships and cooperation between employers and OSHA, between employers and employees, and among employers in the business community.

Small Business Program Example: Anthony Forestry Products

Anthony Forestry Products is a fourth generation, family-owned lumber and wood products company. Its laminated wood products plant in El Dorado, Arkansas employs a staff of 80. The company initiated efforts to improve its safety practices and, in 2001, began working with OSHA's On-Site Consultation Program on a voluntary basis to put in place a working safety and health management system. By 2002, the site was accepted into the SHARP. As a result of this work, the company's workers' compensation loss rate (in losses per $1,000 of payroll) decreased from $18.20 in 1998 to $0.30 in 2007.

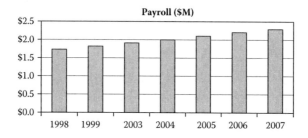

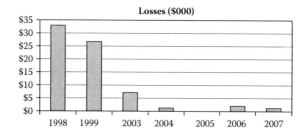

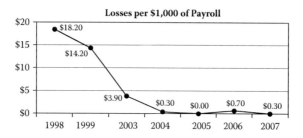

Source: ERG (2008).

Conclusions

- Despite the combined efforts of employers, workers, unions, safety professionals and regulators, more than 4,500 workers lose their lives and more than four million are seriously injured each year. Tens of thousands more die or are incapacitated because of occupational illnesses including many types of cancer and lung disease. The human toll from this loss is incalculable and the economic toll is enormous.
- Many employers in the U.S. have been slow to adopt a workplace "safety culture" that emphasizes planning and carrying out work in the safest way possible.
- Injury and illness prevention programs are based on proven managerial concepts that have been widely used in industry to bring about improvements in quality, environment and safety, and health performance. Effective injury and illness prevention programs emphasize top-level ownership of the program, participation by employees, and a "find and fix" approach to workplace hazards.
- Injury and illness prevention programs need not be resource-intensive and can be adapted to meet the needs of any size organization.

OSHA believes that adoption of injury and illness prevention programs based on simple, sound, proven principles will help millions of U.S. businesses improve their compliance with existing laws and regulations, decrease the incidence of workplace injuries and illnesses, reduce costs (including significant reductions in workers' compensation premiums) and enhance their overall business operations.

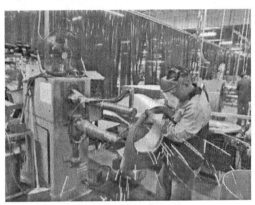

Photo: Roberto Rodriguez, Mesquite, Texas

References

Alsop, P. & LeCouteur, M. (1999). Measurable success from implementing an integrated OHS management system at Manningham City Council. Journal of Occupational Health & Safety – Australia & New Zealand, 15, 565–572.

Angello, J. (2010). A journey to improved safety performance. The Leader, 19(3), 27–29.

ANSI/AIHA (2005). American National Standard – Occupational Health and Safety Management Systems. ANSI/AIHA Z10–2005.

Bunn, W. B. et al. (2001). Health, safety, and productivity in a manufacturing environment. *Journal of Occupational and Environmental Medicine* 43(1), 47–55.

Bureau of Labor Statistics. (2010). Table A-1. Fatal occupational injuries by industry and event or exposure, All United States, 2010. Available at: http://www.bls.gov/iif/oshwc/cfoi/cftb0241.pdf*

Conference Board. (2003). Driving toward "0": Best practices in corporate safety and health.

DOE, (2011). U.S. Department of Energy, Office of Health, Safety and Security. A basic overview of Integrated Safety Management.

European Union, (1989). Council Directive of 12 June 1989 on the introduction of measures to encourage improvements in the safety and health of workers at work. (89/391/EEC).

Gallagher, C. (2001). New directions: Innovative management plus safe place. In W. Pearse, C. Gallagher, & L. Bluff, (eds.) Occupational health & safety management systems: Proceedings of the first national conference (pp. 65–82).

Gallagher, C. et al. (2003). Occupational safety and health management systems in Australia: Barriers to success. *Policy and Practice in Health and Safety* 1 (2), 67–81.

Hasle, P. & Limborg, H. (2006). A review of the literature on preventive occupational health and safety activities in small enterprises. *Industrial Health* 44(1), 6–12.

Huang, Y. H. et al. (2006). Safety climate and self-reported injury: Assessing the mediating role of employee safety control. *Accident Analysis and Prevention* 38(3), 425–33.

Huang, Y. H. et al. (2009). Financial decision-makers' views on safety: What SH&E professionals should know. *Professional Safety* (April), 36–42.

Leigh, J. P. et al. (1997). Occupational injury and illness in the United States: Estimates of costs, morbidity, and mortality. *Annals of Industrial Medicine* 157(14), 1557–1568.

Lewchuk, W., Robb, A., & Walters, V. (1996). The Effectiveness of Bill 70 and Joint Health and Safety Committees in Reducing Injuries in the Workplace: The Case of Ontario. *Canadian Public Policy*, 22, 225–243.

Liberty Mutual Research Institute. (2010). 2010 Liberty Mutual Workplace Safety Index.

Liu, H. et al. (2008). The Pennsylvania Certified Safety Committee Program: An Evaluation of Participation and Effects on Work Injury Rates. RAND Working Paper WR-594-PA.

National Academy of Social Insurance. (2011). Workers' Compensation: Benefits, Coverage, and Costs, 2009.

Ohio Bureau of Workers' Compensation. (2011). Ohio 21(d) SHARP Program Performance Assessment.

OHSAS Project Group. (2007). Occupational health and safety management systems – Requirements. OHSAS 18001:2007. Published as a British Standards Institute standard.

Shannon, H. et al. (1996). Workplace organizational correlates of lost-time accident rates in manufacturing. *American Journal of Industrial Medicine*, 29(3), 258–268.

Shannon, H. et al. (1997). Overview of the relationship between organizational and workplace factors and injury rates. *Safety Science*, 26, 201–217.

Smitha, M.W. et al. (2001). Effect of state workplace safety laws on occupational injury rates. *Journal of Occupational and Environmental Medicine*, 43(12), 1001–1010.

Stanford University. (1981). Improving construction safety performance: The user's role (Technical report No. 260). Department of Civil Engineering.

Steenland, K. et al. (2003). Dying for Work: The Magnitude of US Mortality from Selected Causes of Death Associated with Occupation. *American Journal of Industrial Medicine*, 43, 461–482.

Torp, S. et al. (2000). Systematic health, environment, and safety activities: Do they influence occupational environment, behavior and health? *Occupational Medicine*, 50(5), 326–333.

U.S. Department of Energy, Office of Corporate Safety Analysis. (2009a). Occupational Medicine Newsletter. Corporate safety: Health & safety is working!

U.S. Department of Energy, Office of Health, Safety and Security (HSS). (2009b). A basic overview of the worker safety and health program (10 CFR 851). Outreach & Awareness Series to Advance DOE Mission.

Viscusi, W. & Aldy, J. (2003). The value of a statistical life: A critical review of market estimates throughout the world. *Journal of Risk and Uncertainty*, 27, 5–76.

Yassi, A. (1998). Utilizing Data Systems to Develop and Monitor Occupational Health Programs in a Large Canadian Hospital. *Methods of Information and Medicine*, 37, 125–129.

Footnotes:

Footnote 1: The occupational safety and health community uses various names to describe systematic approaches to reducing injuries and illnesses in the workplace. Consensus and international standards use the term Occupational Health and Safety Management Systems; OSHA currently uses the term Injury and Illness Prevention Programs and others use Safety and Health Programs to describe these types of systems. Regardless of the title, they all systematically address workplace safety and health hazards on an ongoing basis to reduce the extent and severity of work-related injuries and illnesses.

Footnote 2: The "most disabling" injuries are defined by Liberty Mutual as those causing the injured employee to miss six or more days from work.

Footnote 3: For more details see OSHA's Safety and Health Management Systems eTool, available at www.osha.gov/SLTC/etools/safetyhealth/mod1_costs.html.

Footnote 4: If injury and illness prevention programs achieve a 15 percent reduction in injuries and illnesses for employers who do not currently have safety and health programs, the overall reduction in injuries and illnesses for all employers including those that already have programs is estimated at 12.4 percent. Applying this 12.4 percent to NASI's estimate of the $74 billion in direct workers' compensation costs in 2009, workers' compensation savings could be as high as $9 billion per year. With a 35 percent program effectiveness, the overall reduction in injuries and illnesses for all employers is estimated at 30.8 percent and workers' compensation savings could reach $23 billion per year.

Footnote 5: The 15 states are: Arkansas, California, Hawaii, Louisiana, Michigan, Minnesota, Mississippi, Montana, North Carolina, New Hampshire, Nevada, New York, Oregon, Utah, and Washington.

DISCUSSION QUESTIONS

1. What impact does compliance have within the safety and health function?
2. How does I2P2 impact the safety and health function?
3. What areas within the safety and health function are deficient in the area of OSHA standards?
4. Does perfect compliance equate to an exceptional safety program? Why or why not?

3 Value-Based Leadership in Safety

You do not lead by hitting people over the head—that's assault, not leadership.

—**Dwight D. Eisenhower**

The question "Who ought to be boss?" is like asking, "Who ought to be the tenor in the quartet?" Obviously, the man who can sing.

—**Henry Ford**

LEARNING OBJECTIVES

1. Analyze and assess the concept of value-based leadership.
2. Identify and assess the principles of value-based leadership.
3. Analyze and apply the concepts of value-based leadership to the safety and health function.
4. Analyze and assess the leadership qualities and principles within the safety and health function.

As identified by Bob McDonald, president and CEO of Proctor and Gamble, in his "Principles of Value-Based Leadership" presentation to the students at Northwestern University, "It is important for each individual and each organization to get in touch with their education, experiences, culture, family heritage and organizational membership to develop their own set of beliefs."[*] Within the safety and health profession, where the professional can often "be on an island" within the organizational structure, finding and cultivating your individual beliefs and being in touch with your education and experience is especially important for safety and health professionals in order to establish a firm value-based foundation through which to successfully lead the important safety and health function.

The safety and health profession is unique and challenging. The safety and health professional bears many responsibilities in safeguarding each and every employee within the operational workplace. The safety and health professional is also often challenged to balance the inherent conflicts between his or her managerial position within the organization and the employees that he or she is tasked to protect on a

---

[*] "Principles of Value-Based Leadership," presentation at Owen L. Coon Forum, Northwestern University, October 7, 2009. Also see www.kellogg.northwestern.edu/News_Articles/2009.

daily basis. Safety and health professionals are also often tasked with achieving and ensuring compliance with regulatory requirements while also being saddled with economic, labor, and related constraints. This dichotomy in responsibilities can often place safety and health professionals in ethical dilemmas not seen in other professions.

The principles of value-based leadership in safety and health are unique to our profession and significantly different than most management-based professions. Safety and health professionals often perform the safety and health function on a solo basis or in small teams within the managerial structure of the organization for the primary benefit of their employees. Thus, leadership within the safety and health function offers different and specialized challenges requiring the safety and health professional to be in touch with his or her own background and education, experiences and expertise, workplace culture, management structure, family heritage, and organizational memberships through which to develop his or her own moral compass and personalized sets of beliefs through which to lead the safety and health function.

VALUE-BASED LEADERSHIP IN SAFETY AND HEALTH

1. Safety and health leaders must have passion. One trait common among safety and health professionals that is not often seen in other professions is their passion for the job function and profession. Safety and health professions often do not view their function as a job, but rather an advocacy. The safety and health function is one of the few occupations where the safety and health professional's daily activities and programs can have a direct impact on the lives of his or her employees, his or her company, and his or her community. This passion is the basis for and the first principle of our value-based leadership in safety and health.

   A passion for the safety and health function is innate among the professionals in our unique field. This passion, our first value, propels the safety and health professional to go far beyond the norm and the paycheck to strive to achieve perfection. Unlike other professions, when the safety and health professional does not achieve and maintain perfection or as close to perfection as feasible, there is a high probability that an injury or illness may occur.

2. Safety and health leaders must be technically competent. In the safety and health arena, there is no substitute for technical competency. Safety and health professionals, through their formal or informal education and experience, should have a firm grasp on the regulations and standards that form the foundation for compliance. Although safety and health professionals need not know each and every standard verbatim, it is essential that they know where to find the standard or regulation and the method through which to interpret the standard or regulation, and visualize how to design and implement methods through which to achieve and maintain compliance.

3. Safety and health leaders must have a critical eye. Safety and health professionals must have a critical eye and be able to spot potential hazards and issues that others may not see. Safety and health professionals should be able to see situations and issues far in advance and make the necessary

adjustments to ensure the appropriate safeguards are implemented to create and maintain a safe and healthful workplace.

4. Safety and health leaders are made—not born. When you ask a 10-year-old what he or she wants to be when he or she grows up, you hear police person, firefighter, doctor, or other professions that are visible in their lives. How many 10-year-olds have you ever heard say, "I want to be a safety and health professional." Safety and health leaders are made through the acquisition of critical managerial skills and abilities acquired through education and experiences. Think about your own career path. When did you decide to become a safety and health professional? Although many safety and health leaders come to the profession with basic educational background and some inherent leadership abilities, the unique nature of the safety and health profession requires a foundational educational background in the function as well as continuous education throughout their career. In addition to the educational background, safety and health leaders require experience as well as a high-level tool box of managerial skills. Above all, the safety and health leader closely examines and analyzes his or her skills and abilities and has ingrained his or her values and beliefs into his or her leadership of the safety and health function.

5. Safety and health leaders motivate and educate others to lead. One of the first "awakenings" many safety and health professionals encounter is that there are not enough hours in a day for the safety and health professionals to do everything required to develop, maintain, and improve the safety and health function. Safety and health leaders must not only manage the function but also must delegate, educate, and motivate others within the organization to perform vital activities serving the overall function. Safety and health leaders provide the skills and abilities to others to become leaders or champions of the safety and health function and provide support and motivation to ensure their success.

6. Safety and health leaders are change agents. Change is a constant in the safety and health profession due to the constant regulatory additions and modifications, changing workplace environment, changing personnel, and other controllable and uncontrollable factors. Safety and health leaders should not be afraid of change, but should embrace change as a constant and become an agent for positive change in the safety and health environment within their organization. In many circumstances, the safety and health professional is hired by the organization or company with a poor track record in safety and health for the sole purpose of initiating a positive change. Safety and health leaders should accept these challenges and promote positive change that improves the safety and health of the employees in the workplace.

7. Constant intensity. There is often no downtime for safety and health professionals. Unlike other professions, there is always something that needs your attention in the safety and health function. However, the stress that can accumulate within the safety and health professional can be detrimental to both the program and the individual. Safety and health leaders

should assess and evaluate themselves and determine the level at which the stress can be properly managed while maintaining a constant focus and an appropriate level of intensity on the priority issues within the safety and health function.

Given the passion and purpose most, if not all, safety and health leaders possess, safety and health leaders should "pick their battles" and gauge their individual level of intensity in order to maintain a constant and consistent safety and health message within the organization.

This level of intensity may vary depending on the organization; however, a safety and health leader, as part of the overall managerial team, should maintain a clear, consistent, and constant safety and health message to each and every member of the managerial team and all employees.

8. Put the right people in the right places. As noted in the quote by Henry Ford above, placing the right individual in the right job is essential to ensuring the success of the individual as well as the most efficient and effective performance within the job function. Safety and health leaders should carefully assess and analyze the skills, abilities, temperament, and other qualities of the individual as well as the responsibilities, stress, hours, and other factors of the position within the safety and health function. Safety and health leaders should "make the time" to evaluate, correlate, and match the individual to the job, which ultimately will pay dividends over the long run in creating a happier and more successful employee as well as reduced turnover, reduced training costs, and many other cost savings.

Safety and health leaders should strive to create and maintain a diverse workforce where the innovative safety and health ideas are encouraged and supported. Each and every employee has a vested interest in the safety and health function. Encouraging diversity at all levels within the organization creates an environment where safety and health innovation and ownership of the safety and health function can become ingrained within the organization.

9. Integrity and take responsibility for mistakes. The most important trait of a safety and health leader is integrity. Given the myriad issues and challenges that a safety and health leader encounters on a daily basis, mistakes are bound to happen. A safety and health leader will take responsibility for his or her mistakes and not attempt to circumvent the consequences. Safety and health leaders should do what has to be done and not look for the easy way around the issue or situation. Safety and health leaders are often confronted with ethical situations where there is no bright line answers, and the lines can often be blurred between right and wrong. Safety and health leaders have to have analyzed themselves and know their limits and where to draw the line over which they will not cross.

10. Avoid ineffective strategic barriers. Safety and health leaders can often encounter political, policy, procedural, or technical incompetency and other internal and external issues that can hinder progress toward the achievement of the safety and health mission. Safety and health leaders should identify these barriers and circumvent such barriers in order to maintain a clear pathway to the achievement of the proscribed safety and health goals and

objectives. As stated by Sun Tzu in *The Art of War,* "He who knows when he can fight and when he cannot, will be victorious."* Safety and health leaders should assess the barriers to success and select their battles carefully.

11. Forget the politics—ride for the brand. All organizations possess internal politics of varying degrees and natures. It is the nature of the American business, i.e., climbing the corporate ladder. However, within many safety and health functions, the ladder is particularly short. Safety and health leaders should acquire the "lay of the land" within the organization and avoid potential encumbrances that can deter for the achievement of the safety and health mission. The time wasted through internal politics and not on the achievement of the safety and health mission can never be recovered.

    Externally, safety and health leaders should always "ride for the brand." Your organization has entrusted to you confidential information, and the knowledge you have acquired can be potentially damaging to your organization if made public. Safety and health leaders are always professional and should remember to maintain all confidences at all times.

12. See situations from all angles. There are always two sides to every argument, and safety and health leaders should be able to identify and understand both sides of the argument. Safety and health leaders should respectfully listen to all involved before rendering any decision. The safety and health leader's decision-making process should include, in addition to careful listening, a complete and all-encompassing explanation of the reasoning behind the decision. Most individuals, although they may not agree with your decision, will abide by the decision if they are informed as to the reasoning behind it.

13. Self-confidence—make the hard call. Safety and health leaders need to have broad shoulders and a thick skin. Through self-assessment and self-analysis, safety and health leaders should develop their personal moral compass and know where to draw the lines. As a by-product of the development of the safety and health leader's moral compass, self-confidence and the ability to make the hard call will result. Safety and health leadership is not an easy task; however, the safety and health leader will be well prepared to do the right thing.

14. Gratitude and humility. Safety and health leaders should take pride in their work and accomplishments; however, they should acknowledge that most, if not all, accomplishments within the safety and health arena are a team effort. A safety and health leader should be humbled by the team's accomplishments and praise those team members instrumental in achieving the accomplishment. Safety and health leaders have eliminated the word *I* from their vocabulary.

15. Stay on course with your moral compass. The moral compass designed and developed specifically by each safety and health leader should serve as the guiding instrument throughout the safety and health leader's career. There will be ups and downs in the career; however, the moral compass

---

* See *The Art of War,* Sun Tzu, translated by Samuel B. Griffith, Oxford University Press (1963). Also see Sun Tzu quotes at www.brainyquotes.com.

remains consistent. Safety and health leaders should stay the course, following their moral compass throughout their careers. What is right will always remain right, and safety and health leaders should never sell their integrity at any price.

The principles of value-based leadership in safety and health can serve as a road map for safety and health professionals striving to assume a leadership role within the profession or their organization. Although the road map provides direction, it is up to the individual safety and health professional to find his or her "bright lines" through self-assessment and self-analysis. Through this journey of self-analysis and self-assessment, the safety and health professional will learn his or her individual strengths and weaknesses and grow into the role of a true leader. There will be stumbles along the way; however, safety and health leaders simply stand up, dust themselves off, and continue to follow their moral compass on the pathway to success.

DISCUSSION QUESTIONS

1. Prioritize and assess the 15 value-based leadership principles.
2. Identify at least one additional value-based principle that you would include on this list.
3. Where is your moral compass and where do you draw the line? Provide an example.
4. How do your education, experiences, culture, family heritage, and organizational membership assist in developing your moral compass?

4 Empowerment in Safety and Health

I have never been able to conceive how any rational being could propose happiness to himself from the exercise of power over others.

—Thomas Jefferson

Real power has fullness and variety. It is not narrow like lightning, but broad like light. The man who truly and worthily excels in any one line of endeavor might also, under a change of circumstances, have excelled in some other line. Power is a thing of solidity and wholeness.

—Roswell D. Hitchcock

LEARNING OBJECTIVES

1. Analyze and assess the empowerment of employees within the safety and health function.
2. Analyze the methodologies through which to empower employees in safety and health.
3. Identify and assess the action plan to empower employees in safety and health.

Safety doesn't belong to the safety and health professional. The safety and health function, although with different levels of responsibilities, belongs to everyone within the organization. When the organization denotes the safety and health professional to be personally and totally responsible for all aspects of the safety and health function, the safety and health professional is set up for failure or burnout. The safety and health function is substantially too large with far too many "moving parts" for any one person to successfully "keep all of the balls in the air" at all times. The safety and health function requires all employees at all levels in the hierarchy of the organization to be active participants in various aspects of the safety and health function.

The function of safety and health can correlate with the broader structure utilized in team sports. The safety and health professional is the coach. He or she develops the game plan and prepares the players. The team owner, director of personnel, and other administrative functions correlate closely with the management team. The management team's priority is administration, production, and quality. The supervisory level correlates to the team captains on the field and directing the other players. And the employees correlate with the players. If the players know what to do and are empowered and are working together toward a known goal, namely, winning, the team will function at maximum effectiveness. Safety and health, not unlike team sports, takes everyone working together and playing to their potential on each and every play to achieve success.

To get everyone within the organization working as a team in the safety and health arena, there must first be buy-in at all levels of the organization, especially at the top. Each and every level within the organization must accept ownership in the safety and health function. Although many within the organization may not readily accept the additional duties that come with safety and health responsibilities, it is vital to ensure that each and every team member is appropriately prepared to assume these important responsibilities.

Acquisition of buy-in to the concepts of safety and health at all levels is often a challenging process due to the varying priorities at each level within the hierarchy. At the top level of many organizations, the buy-in education often entails more of a financially based concept identifying the costs, return on investment, and manpower requirements. At the mid or managerial level, the buy-in education is more centrally focused on job function, time requirements, and additional duties. At the employee level, the education often focuses on job function, safety and health being everyone's job, safety of fellow employees, personal protective equipment (PPE), and more basic safety and health issues.

Safety and health professionals should consider a systematic and consistent approach wherein each and every team member at all levels is provided the appropriate safety and health tools and education through which to perform his or her specific safety and health activities. With these tools, each and every member of the team will be equipped and educated as to his or her duties and responsibilities within the team structure. Once equipped with the appropriate skills and abilities, each and every responsibility and duty for each position should be fully and completely explained and diagnosed, with each team member being able to analyze, explain, and discuss his or her safety and health responsibilities and duties.

With each level of the organizational hierarchy appropriately prepared, educated, and trained, and with their acceptance of their safety and health duties and responsibilities, mechanisms for appropriate feedback should be instituted. In order for employees as well as managers to be truly empowered in safety and health, appropriate lines of communication with timely feedback should be established. All levels within the organizational hierarchy should be encouraged to provide their thoughts and ideas for safety and health improvement, especially at the managerial and employee levels, and appropriate and timely feedback is essential.

Although the answer acquired to the individual employee's idea may not be what he or she hoped it would be, the simple act of providing timely feedback will encourage participation and support the buy-in to the concept that safety and health belongs to all employees.

With all levels of the organizational hierarchy being truly empowered over the safety and health function, each individual must accept his or her responsibility and duties. Human nature being what it is, acceptance of responsibility requires accountability. The safety and health professional, as the coach of the team, should provide timely and pertinent feedback as to the level of performance and hold each and every member of the organizational team accountable for acceptable safety and health performance.

Although empowerment of the individual team members in safety and health sounds relatively basic in theory, implementation can be substantially challenging. Safety and health professionals should be prepared for a potential backlash from team members who are now being asked to accept new and challenging responsibilities in safety and health. Safety and health professionals should be prepared for managerial team members arguing that they do not have time for additional safety and health responsibilities due to other job responsibilities.

Above all, change in and of itself can be a challenge within any organization. The status quo is being challenged, and team members can no longer point to the safety and health professional as being responsible for safety and health. Each and every team member must now look in the mirror and know that he or she is now responsible for safety and health within the organization, and furthermore that he or she will be held accountable for the achievement of acceptable safety and health performance.

Empowering your employees to take responsibility for safety and health will take time and constant effort. Change in and of itself can be difficult for many individuals within the organization. They have carved out their individual job functions and have been successful for many years. With the tools and education provided, the safety and health professional is now attempting to change the entire safety and health culture within the organization, and these individuals are asked to accept new and different responsibilities in safety and health. Safety and health professionals should be aware that this cultural shift will not happen overnight. Fair, consistent, and constant efforts will be required to achieve the cultural transition.

Safety and health professionals, as the "coach" for the safety and health function within the organization, should be prepared to listen carefully to their newly empowered employees and act upon their recommendations in a timely manner. Although the answer may not be exactly what the individual wishes to hear, it is vital that an answer is provided. A newly empowered employee who enthusiastically embraces his or her new safety and health responsibilities can be quickly throttled if timely feedback is not provided. Right or wrong, a timely response is essential for the newly empowered employee.

For safety and health professionals who may be considering embarking on cultural change within their organization, below an action plan is provided to assist in this process:

1. **Be enthusiastic.** Focus on a positive and winning attitude.
2. **Plan your work and work your plan.** Each and every level of your education, training, and accountability process should be mapped and analyzed. Be creative and challenge the status quo.
3. **Show respect to all employees at all levels.** Who knows the job better than the individual who performs the job daily?
4. **Be accountable.** Your employees, managers, and executives are accountable for their safety and health performance. And the safety and health professional should hold himself or herself accountable. If you say you are going to do something, do it!

5. **Listen intently to your employees.** Safety and health professionals should stop what they are doing and listen intently to any comments or ideas that any employee should bring forward. Safety and health professionals should acknowledge and thank the employee for his or her active involvement in safety and health.

6. **Be a good example.** As the coach, safety and health professionals should always set the standard for safety and health by always wearing the appropriate PPE and always following the safety and health rules. "Do what I say and not what I do" doesn't work in safety and health.

7. **Speak and communicate safety and health daily.** Safety and health professionals should be current on all new regulations, standards, and trends within the function and communicate this information throughout the organization. Safety and health doesn't take a day off; thus, the safety and health professional should be actively communicating and promoting safety and health on a daily basis.

8. **Open your door to employees.** Safety and heath professionals should be readily available for employees to communicate with them. Safety and health professionals should strive to create a level of comfort for all employees to become actively involved and take ownership of the safety and health function.

9. **Be visible on the shop floor.** Although safety and health professionals are often overwhelmed with their many duties and responsibilities, it is important to schedule time each day to be on the shop floor. The visibility of the safety and health professional reinforces the importance of safety and health to employees.

10. **Communicate clearly and concisely.** Safety and health professionals should strive to ensure that all communications are clear, concise, and with no "gray areas" that may be subject to interpretation. Safety and health professionals should ensure that communications are provided at the appropriate educational levels for each function within their organization and with substantial clarity to ensure no ambiguity.

11. **Pat employees on the back.** Safety and health professionals should consistently recognize the safety and health efforts of their employees and provide incentives to spur further involvement and personal ownership in the safety and health function. Incentives do not need to be a formalized program, but activities such as a pat on the back, a handwritten note, or other personalized recognition can be effective.

12. **Avoid taking credit for employees' accomplishments.** Through empowerment of employees, the safety and health function becomes a team effort. Safety and health professionals should always acknowledge their employees' accomplishments in safety and health and should never claim credit for the work of the team.

13. **Leaders without titles.** Safety and health professionals should recognize that many organizations have leaders who function without the requisite title or position. These leaders can be instrumental in the success of your empowerment efforts as well as your overall safety and health program. Safety and

health professionals should recognize these leaders and cultivate their leadership skills and abilities within the overall safety and health function.[*]

14. **Ensure ongoing education and training.** The safety and health function is a marathon rather than a sprint. Ongoing education and training empowers employees with the skills and abilities to take true ownership over the safety and health function.

15. **Acknowledge employee accomplishments.** Continuous and pertinent feedback as to the achievements of the safety and health efforts is essential. Safety and health professionals should recognize and reward the team's efforts within the safety and health function.

16. **Eliminate "I" from your language.** Safety and health professionals should recognize that the ownership of the safety and health function now belongs to the team. The safety and health professional, as the coach, should recognize this transition and always recognize the team's efforts. The simple change from "I" to "we" can go a long way in creating ownership of the safety and health function.

17. **Invite feedback and respond.** Safety and health professionals should create comfortable situations where employees are invited to provide their thoughts and feedback with regards to safety and health issues. Safety and health professionals should recognize that the comfort level and time for acceptance of these new duties and responsibilities may vary among and between individual employees.

18. **Be willing to do the job you are asking them to do.** Safety and health professionals should make the time to spend at least a few minutes at each and every job station within the operation and talk with the employee performing the job function. Safety and health professionals should not ask any employee to perform any task that they themselves would be unwilling to perform.

19. **Be willing to be flexible.** Although safety and health standards can sometimes be restrictive, it is important for safety and health professionals to listen to their team members' ideas and be flexible in their assessments.

20. **Be fair and consistent.** In holding team members accountable for their safety and health responsibilities, safety and health professionals should ensure that they are always fair and consistent in each and every decision.

21. **Be a great coach.** Safety and health professionals, as the coach of the team, should strive to bring out the best safety and health performance in each employee in each job. The safety and health professional should challenge and inspire the team members and encourage top-level safety and health performance.

Safety and health professionals should be aware that this cultural change through empowerment of all levels within your organization can be an elongated process built through day-by-day consistent and constant efforts. The pathway of success is not always smooth, and there are many other variables and outside factors that can delay or derail your efforts. The safety and health professional, as the coach, should guide

---

[*] See Leaders Without Titles, Stephen J. Sampson, HRD Press, Amherst, MA (2011).

and direct, provide positive and negative reinforcement, and be the "guiding light" for the safety and health function. Safety and health belong to everyone; however, team members at all levels need a coach to ensure that they are performing their duties and responsibilities appropriately in order that the team can reach its goals. Once each and every team member is empowered and accepts his or her duties and responsibilities, the environment and culture will transform in a positive manner, creating the ideal safety and health environment for all team members.

DISCUSSION QUESTIONS

1. Identify at least one additional activity through which to empower employees in safety and health.
2. Why is feedback essential within the safety and health function?
3. What do you do if an employee refuses to accept his or her safety and health responsibilities?
4. Compare and contrast positive and negative reinforcement.

5 Impact of Hiring and Discipline in Safety

Hold yourself responsible for a higher standard than anybody else expects of you. Never excuse yourself. Never pity yourself. Be a hard master to yourself—and be lenient to everybody else.

—**Henry Ward Beecher**

I've never known a man worth his salt who in the long run, deep down in his heart, didn't appreciate the grind, the discipline.... I firmly believe that any man's finest hour—this greatest fulfillment to all he holds dear—is that moment when he has worked his heart out in a good cause and lies exhausted on the field of battle victorious.

—**Vince Lombardi**

LEARNING OBJECTIVES

1. Analyze and assess the impact of the hiring process on the safety and health function.
2. Identify and analyze the impacts of disciplinary action on the safety and health function.
3. Identify and analyze the disciplinary action process.

Safety and health professionals are in the business of creating and maintaining a safe and healthful work environment for all employees and minimizing the potential risks for their organization. As previously discussed, each and every team member at all levels of the organization's hierarchy must be actively involved in the safety and health programs and efforts led and directed by the safety and health professional. However, whether it is the team member's job placement, training or lack thereof, the team member's supervisor or coworkers, or simply human nature and the law of probabilities, there will always be some individuals within the organization who will require additional positive as well as negative motivation in order for them to acknowledge, accept, and perform the requisite safety and health duties and responsibilities. The underlying question for safety and health professionals is why.

In my discussions with safety and health professionals over the years, one common thread always emerged, especially with professionals with dual or multiple responsibilities in addition to the safety and health function. This common thread was the amount of time the safety and health professional spent with a relatively small percentage of the overall workforce. Generally speaking, this situation is often generically referred to as the 90-10 percent rule, wherein the safety and health professional spends 90 percent of his or her time with 10 percent of the overall workforce population. Although these percentages may not be truly accurate, safety and health professionals often spend an inordinate amount of their time addressing issues and situations arising from or involving a relatively small number of team members, and these issues or situations often tend to repeat themselves with some frequency. Conversely, because of the time constraints, the safety and health professional often spends a relatively smaller percentage of his or her time with the majority of the team members who function efficiently, effectively, and safely each day without issues.

Although often identified as a human resource function, safety and health professionals should recognize that the selection and hiring process can have a major impact on their safety and health efforts. Far too often, safety and health professionals do not become involved with new employees until the safety and health orientation after the new employee has secured employment with the organization and has been assigned job responsibilities and duties. Given the potential impact on the overall safety and health program, it is important for safety and health professionals to be actively involved in the hiring process and utilize their expertise in selecting, training, and placing new employees in functions through which they can work safely and achieve success.

Generally, the hiring process is managed by the human resource function in most organizations and consists of a job application, interview, assessment, controlled substance testing, background investigation, and reference calls, followed by a formal job offer and acceptance. Depending on the position, there may be additional steps, including, but not limited to, credit checks, medical examination, psychological testing, and other inquiries, depending on the hiring criteria. This process is usually followed by the completion of the required documents with the human resource department, including, but not limited to, tax withholding forms, insurance forms, eligibility to work forms, and numerous other forms. Following this elongated hiring process, the safety and health professional often gets his or her first look at the new employees after they have been formally hired in the proscribed new employee orientation and safety and health training. If the safety and health professional has not been actively involved in the hiring process, it is often too late at the orientation phase to make any type of modification or changes that could create a safer or more healthful job experience for the new employee.

Safety and health professionals should be cognizant of the substantial costs involved in recruiting, screening, hiring, training, and employing a new employee. Generally, the costs fall into the broad category of recruiting expense, training costs, basic salary, employment taxes, benefit cost, space, and equipment costs.* On average, the "estimates range from 1.5× to 3× of salary for the 'fully baked' cost of

---

* See "Starting Up: Practical Advise for Entrepreneurs," Joe Hadzima, Boston Business Journal, www.mit.edu/e-club/hadsima/how-much-does-an-employee-cost.html.

an employee."[*] Given these costs, safety and health professionals should be actively involved in the hiring process to ensure that the optimal candidate is selected and properly trained to perform the specified job function in a safe and healthful manner. Improper selection or improper placement of new employees can create an environment where the new employee will depart the organization, thus incurring further costs, or the new employee may be vulnerable to potential risks that could result in a work related injury or illness. Either way, the organization will incur unnecessary costs that could be avoided through the effective involvement of the safety and health professional in the hiring process.

Utilizing the principles of ergonomics as an example, the manager contacts the human resource department and requests two new employees to perform a lifting function in a new job position on two shifts. The human resources function conducts the hiring process and selects two candidates. Candidate 1 is 6 feet 5 inches in height and will work A shift, and candidate 2 is 5 feet 3 inches in height and will work B shift. Upon placement in the uniform workstation, candidate A will be bending consistently and will be uncomfortable. On B shirt, candidate 2 will be lifting over his or her head consistently and will be uncomfortable. Absent modification to the job and workstation structure, the probabilities exist that either the new employees will leave the job because of the discomfort or incur a work-related injury resulting from the work requirements. Either way, this situation could have been avoided if the safety and health professional was actively involved in the process and was able to address the individual candidate's needs in the workplace. The active involvement in the hiring and placement process could have avoided the costs of replacement of employees when they voluntarily terminate the job, as well as creating a safe and healthful work environment for the new employees that could avoid potential work-related injuries and promote longevity with the organization.

As a caveat, safety and health professionals should be aware and become knowledgeable concerning the numerous laws and regulations that are implicit to the hiring process. Of particular importance to safety and health professionals on the federal level are the Americans with Disabilities Act of 1990[†] and the Americans with Disabilities Amendment Act[‡] addressing disability and requests for reasonable accommodation. Additionally, safety and health professionals should be familiar with the requirements of Title VII of the Civil Rights Act of 1964[§] and the Civil Rights Act of 1991[¶] addressing discrimination in the hiring process and in the workplace. Other laws with applicability to the hiring and employment process includes, but are not limited to, the Equal Pay Act,[**] Pregnancy Discrimination Act of 1978,[††] and Age Discrimination Act of 1967.[‡‡] Safety and health professionals should also

---

[*] See "The Cost of Hiring a New Employee," The Huffington Post, June 4, 2012. Also see www.investopedia. com/financial-edge/0711/the-cost-of-hiring-a-new employee.html.
[†] 42 USC Section 12101 (Pub. L. 101-136).
[‡] Pub. L. 110-325.
[§] Pub. L. 88-352.
[¶] Pub. L. 1102-166.
[**] 29 USC Section 206(D).
[††] See Section 701 of the Civil Rights Act of 1964.
[‡‡] 29 USC 621 (Pub. L. 90-202).

be aware that there are numerous state and local laws and regulations applicable to the hiring and employment process, as well as case law creating requirements and exceptions to established laws.

Selecting and placing the right individual in the right job and treating the individual right will create a working environment where the individual will flourish as well as the organization. Individuals who are happy in their work activities are often more productive, work safer, and create an environment that promotes longevity and creativity. Conversely, individuals who have been placed in jobs that have physical, educational, or psychological demands beyond their capabilities will wither in the job, potentially creating a higher risk for injury, voluntary departure from the organization, thus incurring replacement costs, or other issues detrimental to the organization.

Enforcement of proscribed disciplinary measures for violation of established safety and health policies and procedures are essential for an effective and efficient safety and health program. Although positive reinforcement has been recognized as being more effective, the negative reinforcement provided by an established and strictly enforced disciplinary procedure and process ensures that safety and health policies, procedures, and programs are being followed. Most organizations use varying combinations of positive and negative reinforcement for enforcement of organizational policies ranging from attendance to work performance. However, although enforcement processes are in place, some organizations' enforcement of safety policies and procedures can often be viewed and enforced differently than enforcement of production and quality related areas. Fair and consistent enforcement utilizing both positive and negative reinforcement is essential for an effective and efficient safety and health program.

From a positive reinforcement prospective, many organizations utilize coaching and counseling as part of the overall disciplinary system. Given the cost in terms of dollars to the organization and the impact on the individual, it is important to exhaust all avenues in positively and negatively motivating individuals before utilizing the "workplace death penalty" of involuntary termination. Conceptually within the safety and health function, positive and negative reinforcement is utilized simply to improve safety and health performance. Coaching is often the first step in the progressive disciplinary system established by many organizations. Coaching within the safety and health function, not unlike athletic coaching, is a positive reinforcement method through which to provide guidance and skills to improve the individual's safety and health performance. Counseling is often the next step in the disciplinary progression wherein an individual has erred in some manner within the safety and health function and individualized positive instruction is provided to the individual to correct the situation or issue. Both coaching and counseling are methods through which unacceptable safety and health actions, inactions, behaviors, or skills can be immediately corrected through the intervention and providing of positive guidance.

Virtually all organizations have a method through which to offer negative reinforcement to individuals who have not corrected the unacceptable safety and health performance through coaching or counseling. Many organizations use a progressive disciplinary system consisting of a number steps with penalties increasing in severity, ultimately cumulating in involuntary termination from the workplace. Progressive disciplinary systems often utilize a verbal warning stage followed by

a written warning in the individual's personnel file. If the unacceptable performance or behavior has not been corrected and continues by the individual, the more
severe reinforcements of suspension from work and thus a reduction in pay follows.
If the organization's efforts have failed throughout the coaching, counseling, verbal
warning, written warning, and finally suspension, involuntary termination from
employment is the next and final step within the disciplinary progression. At this
point, the individual has been provided a substantial number of opportunities to
change his or her safety and health actions or behavior, and the organization can no
longer continue to assume the risk of the individual in the workplace. The individual
loses his or her employment, but the organization loses in terms of skills and costs
invested in the individual as well as replacement costs.

When enforcing safety and health policies and procedures, professionals should
have a firm grasp of the organization's progressive disciplinary procedures. Conceptually, safety and health professionals should be the leaders, the consultants,
and the "eyes and ears" on the production floor. Disciplinary actions for safety and
health violations should be conducted by the individual's supervisor and manager in
conjunction with the human resource professionals. However, with this being said,
safety and health professionals often are required to initiate disciplinary actions,
especially when the violation of safety and health procedures or policy can create
imminent danger for the individual, fellow employees, or the operations. When
issuing disciplinary action for an employee, safety and health professionals should
consider the following:

1. Disciplinary action at any level can be a legal minefield. Know the status of
   the individual before issuing disciplinary action.
2. Documentation is essential! If it is not documented, you don't have it!
3. Know your organization's policies and procedures inside and out.
4. Never get upset. Always calm down before issuing disciplinary action.
5. Always investigate and contemplate before issuing disciplinary action.
6. Provide disciplinary action in person (rather than by letter or email).
7. Be prepared with all documentation before meeting with the individual.
   Remember, after several positive and negative reinforcement levels, the
   individual knows what this disciplinary meeting is all about.
8. Always have a witness in the room during disciplinary meetings.
9. Protect yourself and your organization before, during, and after the disciplinary meeting. Individuals can become upset during disciplinary meetings.
10. Be prepared to address posttermination issues such as workers' compensa-
    tion claims. Human resources can usually assist with questions regarding
    401K plans, COBRA, or other benefits.

A relatively new phenomenon safety and health professionals should be aware
is that of social networking and covert videotaping. Safety and health professionals
should always act professionally and never make promises or statements that can be
misconstrued during a disciplinary meeting. If the safety and health professional is
nervous, it may be advisable to script the statements to be made to ensure complete
and total accuracy. In today's Internet society, safety and health professionals should

anticipate discussion by the individual or others within social network websites regarding the disciplinary action, as well as possible videotaping of the disciplinary issues, general safety and health conditions, and related issues on the Internet.

In summation, each and every individual hired to perform work within your organization should be properly prepared, trained, and positioned to be able to work safely and achieve success. Individuals usually do not fail—the organization fails the individual. Safety and health professionals should be actively involved in the selection, hiring, and training processes in order to equip the individual with the safety and health skills to be successful. When an employee is not properly prepared, it is essential that the safety and health professional utilize the tools within the progressive disciplinary system to positively motivate the individual to work in a safe and healthful manner. Where the positive reinforcement is not effective, it is essential the safety and health professional, in conjunction with the organizational team, provide fair, consistent, and appropriate negative enforcement to motivate the individual to work safely and healthfully. When the individual just "doesn't get it" to work safely after substantial positive and negative reinforcement steps, it is essential that the safety and health professional and organization follow the disciplinary procedure to remove the individual from the workplace for his or her own safety and the safety of his or her fellow employees.

DISCUSSION QUESTIONS

1. Is disciplinary action important to the safety and health function? Why or why not?
2. Please describe in detail your company or organization's disciplinary policy/procedure.
3. How does your hiring process impact the safety and health function?
4. Identify at least one pitfall of the disciplinary process.

6 Don't Be Afraid to Fail

Seven National Crimes:

1. I don't think.
2. I don't know
3. I don't care.
4. I am too busy.
5. I leave well enough alone.
6. I have no time to read and find one.
7. I am not interested.

—William H. Boetcker

Giving up is the ultimate tragedy.

—Robert J. Donovan

LEARNING OBJECTIVES

1. Assess and analyze change within the safety and health function.
2. Analyze how OSHA standards integrate into the safety and health function.
3. Assess and analyze the use of a variance within the safety and health function.

Change in the workplace is constant. Technology is upgraded. Hardware and software change. Employees change jobs. Companies are bought and sold, creating new internal and external structures. Job functions are merged and modified. New products are created and the processes through which they are manufactured are installed. Safety and health professionals should be prepared for change and be able to adapt and try new and innovative ideas through which to safeguard their employees and eliminate or minimize the potential risks in their workplace.

Safety and health professionals should be prepared to adapt and innovate to address the workplace changes. This often means going beyond the standards and experimenting with new concepts, ideas, and programs to safeguard their employees. Although the Occupational Safety and Health Administration has provided a substantial number of standards addressing a broad scope of workplace hazards, the Occupational Safety and Health standards provide only the foundation through which safety and health professionals can build and adapt to create a safe and healthful environment in your specific work environment.

The foundational requirements within the safety and health profession are the standards proscribed by the Occupational Safety and Health Administration. These standards are mandatory in nature and are promulgated by the Occupational Safety and Health Administration. This myriad of standards (where federal or state plan) offer a wide spectrum of requirements to address general as well as industry- specific hazards in the workplace. In general, the standards are categorized as horizontal, requiring a substantial portion of the general industry or construction industries to achieve and maintain compliance. Such standards include the control of hazardous energy (LOTO)[*] and bloodborne pathogens.[†] The vertical standards are usually industry specific and are often developed to address a specific industry hazard. It is often the safety and health professionals' responsibility to identify and correlate these requirements with the specific hazards identified in the workplace and ensure compliance is achieved and maintained with each and every applicable standard. Standards that are not applicable to the specific workplace do not need to be addressed by the safety and health professional.

Even with the multitude of standards promulgated by the Occupational Safety and Health Administration, simply achieving and maintaining compliance with the applicable standards provides only the foundational elements for your safety and health program; however, the safety and health professional should build upon this foundation to address unique risks and hazards in the workplace that are not addressed in the standards. Additionally, safety and health professionals should be exploring other methods through which to motivate and reinforce safety and health within the workforce. Safety and health professionals should be aware that the promulgation process to acquire a standard often takes a substantial period of time, and often the technology, processes, and correlating risks being created are not specifically addressed within a specific standard. Although the Occupational Safety and Health Administration does address risks and hazards beyond the specific standards through the general duty clause,[‡] the burden to create and maintain a safe and healthful work environment has been shifted to the safety and health professional and his or her organization to address risks and hazards that do not possess a specific standard. For example, there are few standards addressing the specific risks and hazards involved in nanotechnology. However, safety and health professionals working in organizations utilizing this technology should address the potential risks and hazards and develop appropriate safeguards to protect their employees. To achieve these safeguards, safety and health professionals must utilize their knowledge and ingenuity to develop programs and safeguards to minimize or eliminate the risks. Often, with minimal guidance from the standards, safety and health professionals must try new and innovative ideas, and not every idea is an overwhelming success. It is essential that safety and health professionals continue to strive to achieve and maintain the safest and most healthful work environment possible and not be afraid to try new and innovative ideas to improve the workplace. With innovation often comes failure. So long as employees are appropriately protected, safety and health professionals should not be afraid to fail in their endeavors to improve the safety and

---

[*] 29 CFR 1910.147.
[†] 29 CFR 1910.1030.
[‡] 29 CFR 1919.5(a)1.

health of their workplace. Appropriate evaluation, assessment, experimentation, and protections are essential; however, safety and health professionals can learn as much from a failure as a success.

Safety and health professionals should consider utilizing all appropriate sources when addressing risks and hazards that may be beyond the standards and the norm. In addition to the network of industry-specific safety and health professionals and consultants, consideration can be provided to governmental avenues, such as the National Institute for Occupational Safety and Health (NIOSH), and safety and health professional organizations, such as the National Safety Management Society. In many states, the Occupational Safety and Health Administration or correlating state plan program may offer assistance through its education and training programs.

A more formalized method through which to address risks and hazards beyond or in conflict with the standards is a variance. As can be seen from the application process for a variance in this chapter, a variance request can provide the safety and health professional the opportunity and avenue through which to assess, evaluate, and offer new and innovative methods through which to safeguard employees in situations beyond the standards or where the operational structure conflicts with the achievement of compliance with a specific standard. Safety and health professionals should evaluate and assess methodologies through which to safeguard their employees in the specific situation and codify and be ready to support the creation of a work environment that is safe and healthful through the variance application process. In addition to the substantial scrutiny and assessment provided by the application process, safety and health professionals can be assured that their proposed idea(s) identified in the application will be exempt from possible violation in future inspections, if approved.

The positive motivation of each and every employee to work and think safely and in a healthful manner on a daily basis has always been a priority for safety and health professionals. Throughout the decades, safety and health professionals have utilized a number of ideas and incentives through which to achieve this motivational level within their workforce. Due to the relative success of such efforts, commercially available incentive programs are now readily available in a variety of products and price ranges. However, safety and health professionals should be aware that incentive programs should not be utilized as a substitute for base-level compliance and behavior-based safety and health programs, and should be utilized for their intended purpose only. Many motivational and incentive-type programs are the ideas of safety and health professionals and are unique to the workforce within their operations.

When compliance is achieved and being consistently maintained, the foundational elements of your safety and health programs are operationally sound, and the safety and health professional is exploring methods through which to change the mindset and culture in a positive manner toward safety and health, there should be no fear in pursuing new and positive ways through which to motivate employees. Although there is often a monetary cost as well as time commitment by the organization, and there is no guarantee of success, pursuit of new and innovative ideas through which to motivate employees to work safely and healthfully can be a positive step in the improvement of your overall safety and health program.

How to Apply for a Variance

OSHA has no single, uniform application form for an employer to apply for a variance. *Therefore, to apply for a variance, an employer must review the specific regulations applicable to each type of variance, and submit the required information.* Generally, the application can be in the form of a letter with the following information included:

·   *An explicit request for a variance*

·   **The specific standard from which the employer is seeking the variance.**

·   *Whether the employer is applying for a permanent, temporary, experimental, national defense, or recordkeeping variance, and an interim order. (If the application is for a temporary variance, state when the employer will be able to comply with the OSHA standard.)*

·   *Describe the alternative means of compliance with the standard from which the applicant is seeking the variance. **The statement must contain sufficient detail to support, by a** preponderance of the evidence, a conclusion that the employer's proposed alternate methods, conditions, practices, operations, or processes would provide workers with **protection that is at least equivalent to the protection afforded to them by the standard** from which the employer is seeking the variance. (National defense variances do not require such a statement, and the statement submitted by an employer applying for a temporary variance must demonstrate that the employer is taking all available steps to safeguard workers.)*

·   *Provide the employer's address, as well as the site location(s) that the variance will cover.*

·   **A certification that the employer notified employees using the methods specified in** the appropriate variance regulation.

·   *An original copy of the completed variance application signed by the employer or an authorized representative of the employer.*

Submit the original of the completed application, as well as other relevant documents[i], to:

By regular mail:

Assistant Secretary for Occupational Safety and Health
Director Office of Technical Programs and Coordination Activities
Occupational Safety and Health Administration
U.S. Department of Labor
Room N3655
200 Constitution Avenue, NW
Washington, DC 20210

By facsimile:

202 693-1644

Electronic (email):

VarianceProgram@dol.gov

Experience in processing variance applications indicates that such applications are not appropriate in the following situations:

- *The variance is from a "performance" standard, i.e., a standard that does not describe a specific method for meeting the requirements of the standard.*

- *The variance is from a "definition" in a standard, i.e., a provision that defines a term used in the standard, but does not expressly specify an action for meeting a requirement of the standard.*

- *The variance is a request for review and approval of a design or product developed for manufacture and commercial use.*

- *There is an OSHA standard in effect that allows the requested alternative.*

- *There exists an OSHA interpretation that permits the requested alternative.*

- *There is an updated edition of a national consensus or industry standard referenced in the OSHA standard, and that is the subject of the variance application, that permits the requested alternative.*

- *The application requests an exemption or exception from the requirements of the standard.*

- *If the application is for a temporary variance, the employer applied on or after the date the standard became effective.*

- *The applicant is contesting a citation involving the standard in question, or has an unresolved citation relating to this standard.*

- *The application involves locations that are solely within states or territories with OSHA-approved plans.*

- *The application is from a Federal agency.*

Variance Application-Related Information

All information or documents submitted in an application becomes public unless the employer claims that some of the material consists of trade secrets[2] or confidential business information.[3] Employers seeking protection of trade secrets or other confidential business information in their variance application and supporting documents must include a request for such protection, as well as a justification for this request, in their variance application. Employers requesting such consideration should note that OSHA will assess the variance request solely on the basis of information that is available to the public.

Variance Application Checklists

To increase transparency and accessibility of variance related information, OSHA developed Variance Application Checklists designed to assist variance applicants to determine if their application for a variance is complete and appropriate.

- Permanent Variance Application Checklist [27 KB PDF*, 2 pages]
- Temporary Variance Application Checklist [27 KB PDF*, 2 pages]
- National Defense Variance Application Checklist [53 KB PDF*, 1 page]
- Experimental Variance Application Checklist [30 KB PDF*, 3 pages]

Variance Application Forms

The Variance Application forms are designed to assist prospective variance applicants to understand what information is required for a variance to be granted in a straight-forward, effective and user-friendly manner. Use of the Variance Application forms coupled with the Variance Application Checklists significantly reduces the burden of wading through the complexity of Federal Standards in order to interpret and understand the information obligations associated with applying for a variance.

- Permanent Variance Application Form [58 KB PDF*, 6 pages]
- Temporary Variance Application Form [62 KB PDF*, 7 pages]
- National Defense Variance Application Form [53 KB PDF*, 7 pages]
- Experimental Variance Application Form [60 KB PDF*, 6 pages]

Site Assessments

Either staff from OTPCA or the OSHA area office staff will perform an assessment of the employer's worksite when deemed necessary. OSHA will conduct site assessments when making decisions regarding the adequacy of an application and it needs further information to process an application. Site assessments are especially useful for temporary or experimental variances, or when OSHA receives employee complaints regarding a variance application. Generally, experimental variance applications will necessitate an onsite assessment to verify that the proposed experimental conditions are safe and healthful for workers. For temporary variances, the site assessment would investigate the availability of appropriate practices, means, methods, operations, and processes needed to come into compliance with the standard, as well as the ability of the employer to meet specific deadlines.

OSHA will arrange the site assessment with the employer in advance of its arrival. There are three parts to the subsequent site assessment: the opening conference, the site investigation, and the closing conference. Site assessments are not compliance inspections, and the OSHA compliance safety and health officers (CSHO) participating in a site assessment will not issue citations to the employer. However, the CSHO will inform the employer of any imminent dangers observed, and will request the employer to abate the hazard; if the employer refuses to do so, the CSHO will inform the nearest OSHA area office of the danger, and the area office will issue a citation. In addition, the site-assessment team will inform the employer of other hazards observed, and the need to abate the hazards, but will not issue citations for these hazards nor inform the area office of the hazards.

Granted or Denied Variances

After an employer submits a variance application, OSHA can either grant or deny the variance. Prior to granting a variance or interim order, OTPCA coordinates a thorough administrative and technical review of the application with other directorates and offices in OSHA, as well as its regional and area offices when appropriate. For permanent and experimental variances, the technical review determines if the alternate method proposed by the employer affords workers protection that is as effective as the protection that would result from complying with the standard from which the employer is seeking a variance. OSHA will deny a variance if the variance application fails the administrative or technical review process, including a failure to demonstrate that the proposed alternative would protect the employer's workers at least as effectively as the standard from which the employer is seeking the variance.

Copies of granted and denied variances, as well as interim orders, are available by accessing links from this Variance Web page.[4]

---

[4] OSHA website: www.osha.gov.

Safety and health professionals and their designed programs, policies, and procedures are failures on a daily basis. Despite the safety and health professional's extraordinary efforts, employees continue to be injured and killed in the American workplace on a daily basis. Safety and health professionals should not fear failure but utilize each failure as a learning experience through which to strengthen their safety and health efforts. Fear can be paralyzing, and often safety and health professionals are challenged at every turn. It is often easier to do nothing, but in safety and health, doing nothing is not a viable course of action. Don't be afraid to take on new challenges, try new ideas, and consistently work to create and maintain the safest and most healthful work environment possible for your employees.

DISCUSSION QUESTIONS

1. Identify one failure in your life and explain how you were able to learn from the experience.
2. Discuss the possible use of a variance action and provide an example.
3. Identify one area in which there is no OSHA standard but one is needed.

7 Rethinking Workers' Compensation

To industry, nothing is impossible.

—Latin proverb

Ideas must work through the brains and the arms of good and brave men, or they are no better than dreams.

—Ralph Waldo Emerson

LEARNING OBJECTIVES

1. Assess and analyze how accidents happen.
2. Analyze and identify the components of the workers' compensation process.
3. Acquire an understanding of the Occupational Safety and Health Administration (OSHA) recordkeeping process.
4. Identify the strengths and weaknesses of the current workers' compensation system.

Where our safety and health efforts fail, there is often an injury or illness incurred by our employees. Work-related injuries and illnesses, in addition to the pain and suffering by the injured or ill employee, cost organizations money. The goal of the safety and health professional is to eliminate or minimize the potential risks in the workplace in order to avoid the potential hazards that can result in workplace injuries and illnesses. The reason most organizations employ a safety and health professional on staff is primarily and fundamentally because work-related injuries and illnesses cost the organization money. If there was not a monetary cost correlated to work-related injuries and illnesses, many organizations would simply marginalize the compliance and related requirements and align safety and health responsibilities within another organizational function.

As identified by many safety and health professionals over the years, work-related accidents do not simply happen. Work-related accidents are the result of identifiable hazards in the workplace, workplace activities and stressors, deficiencies in training and education, and related factors. As identified by C. Everett Marcum in his "domino sequence," the risk of accidents is the result of inadequate task preparation, substandard performance, and miscompensated risks.[*] If the risk cannot be eliminated or

---

[*] Modern Safety Management, C. Everett Marcum, W V Worldwide Safety Institute (1978).

minimized, the accident will happen. Virtually everything that happens following the accident and subsequent injury or illness is negative for the organization and costs the organization money. If the risk is eliminated and the accident does not happen, there is no cost to the organization. Safety and health professionals know that proactive safety measures and programs prevent accidents, and thus provide an exceptional monetary return on investment for progressive organizations as well as for personal benefits for employees.

When an employee is injured on the job, safety and health professionals should be knowledgeable with regards to the numerous and substantially different laws, rules, and regulations, as well as federal and state agencies that become involved in the situation. Starting with workers' compensation, safety and health professionals should be aware that each individual state possesses its own workers' compensation administrative system with individual state laws governing every aspect of workers' compensation within the state. Unlike OSHA, there is no overriding federal oversight for state workers' compensation programs. Generally, most workers' compensation systems have been designed to provide payment for medical, time loss from work, and a monetary value on the loss of function or appendage to the injured employee. Each state establishes the values, and the workers' compensation system is funded by employers and state revenues. Each state possesses individual required forms, documents, and procedures.

Workers' compensation evolved as a compromised method to eliminate the burden on the court systems and to provide certainty for both the employee and the employer. For the injured employee, the benefits of workers' compensation include payment of all medical costs, a quick payment of time loss benefits, monetary payment for permanent losses, and the relative ease through which to acquire benefits through an established administrative process. For the employer, who is usually required to participate in the state workers' compensation system, the major benefits are that injured employees would be barred from pursuing legal action, which could result in higher costs, and the certainty of the cost as established by the individual state workers' compensation laws. For the courts, the volume of workers' compensation cases would be managed through the established administrative system, and most avoid the court system altogether. Generally, the negative aspects of workers' compensation for injured workers include reduced benefits, such as 66% of their average weekly wage for time loss benefits, a fixed schedule for permanent or total disability, and often the bureaucracy of the workers' compensation system. For employers, the primary issue often involves the increasing costs of workers' compensation, usually established by state legislatures, as well as the management of the workers' compensation claims. Although the workers' compensation systems in most states have all but eliminated the need for the courts to be involved in workers' compensation claims, new theories and legislation have created inroads wherein certain claims or actions would be permitted to be heard before the state court system.

For safety and health professionals, the required workers' compensation coverage acquired by employers can take different administrative forms, including self-insurance, self-insured with outside administration, or traditional coverage. If self-insured, the cost of the workers' compensation claim is paid directly by the employer, and often large bonds are required by the state workers' compensation agency to ensure against default. This option is usually utilized by very large organizations, and the claims management can be internal or through a third party. Traditional coverage can be through a third party or often through a quasi-governmental entity that manages the claims. Under this type of system, there is often a 3- to 5-year mode system where improvements in safety and health programs with the correlating reduction in claims are evaluated on a 3- to 5-year period of time.

With individual states establishing and effectively managing their workers' compensation system, safety and health professionals should be aware that there are many similarities; however, there are also many differences among and between individual state workers' compensation systems. Absent one system to preside over workers' compensation in the United States, an expanding scope of inequities has developed among and between state programs. For example, the benefit levels for time loss and permanent partial disability for identical injuries in different states could be substantially different in terms of monetary payments to the injured employee. Safety and health professionals should be aware of the differences among and between the various state workers' compensation systems in which their operations are located, and be aware that the systems are not "apples to apples" when comparing costs for work-related injury and illness claims. Additionally, safety and health professionals should also be aware that any and all work-related injury or illness will generate a claim that is the mechanism through which to pay the medical providers and other benefits. A claim equals costs.

Under the recordkeeping requirements established by the Occupational Safety and Health Administration and adopted by virtually all of the state plans, the work-related injury or illness is required to be recorded on the OSHA provided forms. However, safety and health professionals should be aware that the criteria through which a work-related injury or illness becomes recordable is substantially narrower than that of a workers' compensation claim. A work-related workers' compensation claim in which monies are spent by the employer may not meet the criteria to become an OSHA recordable case. Thus, given the different criteria between workers' compensation and OSHA, the same injury or illness could potentially be a difference, depending on which system is being evaluated.

How do you decide whether a particular injury or illness is recordable? The decision tree for recording work-related injuries and illnesses shows the steps involved in making this determination.

For safety and health professionals, this can be a bit confusing. The same work-related injury or illness in operations in different states will generate a claim; however, the monetary cost and results of the claims can be substantially different. The same work-related injury or illness may not meet the threshold and criteria to be recordable under the OSHA recordkeeping system, and thus may not be identified under the OSHA system. Additionally, the OSHA recordkeeping standard places the burden upon the employer to ensure accuracy with the criteria, however subjective, wherein the workers' compensation system is self-regulating given the demand for payment for medical services rendered by the health care provider. For example, if an employee goes to the emergency room for treatment, one of the first questions the health care provider asks is whether this is work related. If the answer is yes, the claim is generated so the emergency room can acquire payment for its services.

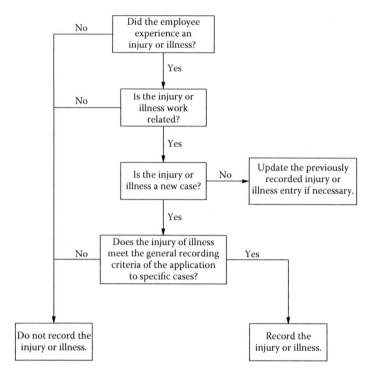

FIGURE 7.1   Chart demonstrating whether or not to record an injury or illness.
29 CFR 1904.4(b)(2).

There is a cost to the employer or organization. Conversely, the employee may have generated a claim and has been treated by the emergency room; however, the diagnosis does not meet the recordkeeping criteria and thus would not be a recordable injury or illness under OSHA recordkeeping criteria (Figure 7.1). There would still be a cost to the company or organization due to the filing of the claim and subsequent medical treatment.

For employers with many operations in multiple states, this multiagency system with specified federal and state laws, requirements, and regulations can be substantially complex to manage. Given the complexity and duplicative costs, have these independent but correlating systems placed the cart before the horse? Is there a need for 50+ independent state agencies administering workers' compensation with different rates and rules? Would it be cost beneficial and more equitable in terms of benefits if there was one agency at the federal level to manage workers' compensation for all work-related injuries and illnesses on a national level? Should additional funding be provided to efforts to prevent the occupational injury or illness from occurring rather than utilized in a reactive manner after the work-related injury or illness has occurred? Remember, if the work-related injury or illness is prevented, there will be no workers' compensation cost. Is there a better way to manage safety and health as well as workers' compensation?

Utilizing the concept of reducing or eliminating the risk, and thus the work-related injury or illness as well as the correlating costs, who are the responsible individuals within the organization or company, and what qualified these individuals to assume this responsibility? What qualifies a person to be a safety and health professional? Are there any testing, licensing, and guidance for professional behavior or enforcement within the safety and health profession? At this point in time, there are no specific requirements or qualifications for an individual to be considered a safety and health professional. There are degree programs, certification programs, voluntary association designations, and other training; however, there is no required testing or licensure required for safety and health professionals as with other professions, such as medicine and law. Companies and organizations are entrusting the safety and health of their employees as well as the management of the inherent risks of their operations to individuals with varying skills, abilities, and experiences, and often have no way to evaluate their skills and abilities.

As with other professions, there is no mandatory code of professional ethics or conduct that is required when an individual holds himself or herself out as a safety and health professional. Although there are many safety and health organizations with codes of professional conduct, their codes are voluntary and only enforceable on their membership. Individual companies or organizations often have codes of professional conduct, but again, they are only enforceable on their employees. At this point in time, there is no testing, licensing, or code of professional conduct that is mandatory for an individual to consider himself or herself a safety and health professional. Given the fact that the safety and health of many are in the hands of the safety and health professional, should there not be a testing and licensing process, similar to the medical and legal professions, where minimum standards are established and verified, as well as mandatory enforcement of professional standards and conduct?

As future safety and health professionals, it is up to you to rethink our current systems and initiate changes that would be beneficial to all. If the work-related injury or illness is prevented, there is little or no cost. Why do we spend far less on prevention than we do on workers' compensation? Should each individual state possess its own workers' compensation system with varying rules, regulations, and benefits? Should OSHA possess a different system to identify and calculate work-related injuries and illnesses? And lastly, should there be a minimum level of qualification within our profession and a method through which to enforce acceptable behavior within the profession? The safety and health profession has been emerging over the past decades, and it's up to you to take the safety and health profession to the next level.

NIOSH Center for Workers' Compensation Studies (CWCS)

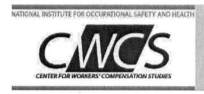

In 2013,the NIOSH Center for Workers' Compensation Studies (CWCS) was created to organize workers' compensation data that are already being analyzed by NIOSH researchers in existing programs, such as the <u>Economics Program</u>, <u>Surveillance Program</u>, and the <u>Center for Motor Vehicle Safety</u>. This is important because coordinated workers' compensation research has been conducted mostly at large commercial insurers, state-based insurers, or organizations such as the <u>National Council on Compensation Insurance (NCCI)</u>, which cannot always widely distribute their research findings.

We are conducting analyses across a wide range of industry sectors, including:

- *Construction*
- *Healthcare and social assistance*
- *Manufacturing*
- *Public safety*
- *Services*
- *Mining*
- *Transportation*
- *Warehousing*
- *Wholesale and retail trade*

The mission of the CWCS is to use workers' compensation data to prevent and reduce the severity of workplace injuries and illnesses. We will work with public and private partners to maximize the use of their own workers' compensation data.

On this page, you will learn about our current research, publications, and related resources. For more information about the Center, see links below or please contact the CWCS Director, <u>Steve Wurzelbacher.</u>
<u>*What is workers' compensation?*</u>
<u>*Goals*</u>
<u>*Current Research Projects*</u>
<u>*Publications*</u>
<u>*Related Resources*</u>

- <u>*National Institute for Occupational Safety and Health (NIOSH)*</u>
- *Centers for Disease Control and Prevention*

DISCUSSION QUESTIONS

1. What is the difference between OSHA recordkeeping and the workers' compensation process?
2. What is a recordable injury or illness?
3. How do accidents happen, and what is the related cost?
4. What are the current licensure requirements, testing procedures, and code of professional conduct in the safety profession?

8 Creating Safety Creativity in the Workplace

The first step in handling anything is gaining the ability to face it.

—L. Ron Hubbard

One cannot manage too many affairs; like pumpkins in the water, one pops up while you try to hold down the others.

—Chinese proverb

LEARNING OBJECTIVES

1. Analyze and assess methods to energize employees within the safety and health function.
2. Understand and analyze Maslow's principles.
3. Understand how to energize and empower employees in safety and health.

The safety and health function is in the unique position of having a product that everyone wants and everyone needs. From the top to the bottom in every organization, virtually every employee at every level will say that safety and health are very important and a priority. However, although employees say safety and health are important to them, the day-to-day activities and the continued resulting work-related injuries and illnesses say otherwise. Management often states that safety and health are priorities, however provides minimal financial and other support. It is our responsibility as safety and health professionals to ingrain and empower each and every employee with the skills and abilities to perform their job in a safe and healthful manner, as well as motivate each employee to take possession of his or her own safety and health, as well as the safety and health of his or her fellow employees, in order to truly achieve the cultural shift necessary within the workplace.

In most organizations, the safety and health professional is the chief designer of the safety and health programs, as well as trainer, purchaser, investigator, motivator, and any and all other functions necessary to move the safety and health efforts forward. As the chief architect, safety and health professionals should look for new and innovative methods to integrate each and every employee into the safety and health function. And above all, safety and health professionals should explore ways to energize and invigorate their safety and health efforts. Safety and health activities do not need to be boring! And there is no problem having fun when learning and performing safety and health training and other activities.

Safety creativity can be cultivated through empowerment and education of your employees. Employees must be provided the safety and health tools, skills, and education to assemble the foundation through which to base their creativity and ingenuity. This foundation permits each employee to have a firm grasp of his or her safety responsibilities and duties, thus allowing the employee to think and explore beyond the basic aspects of his or her job function. As identified in Maslow's hierarchy of needs,[*] your employees should, at a minimum, have achieved the physiological and safety levels within the Maslow's hierarchy before their creativity can flow. Within the physiological level or basic survival level, this can include, in addition to the survival basics, the financial aspects and workplace environment under which your employees are working. For example, if your employees are working two jobs and simply trying to make ends meet, the probability of the employees having time to think creatively within the job function is relatively low. From the elevated safety level, if the workplace is not a safe and secure environment, employees will spend more of their time on personal safety and security, and the probability that the employee will allot time to think creatively are minimal.

However, if Maslow's hierarchy of needs,[†] including the physiological level, safety level, and belonging level, is present in the current environment, employees should gradually feel more confident in the safety arena and achieve a level of self-esteem and respect for themselves and their fellow employees. As this confidence grows, employees will move to the self-actualization level within Maslow's hierarchy,[‡] where employees' spontaneity and creativity within the safety function will emerge. Safety and health professionals should strive to create and maintain an environment where employees' basic needs are addressed and employees can grow and prosper within the safety function.

In order to prepare your employees to engage their safety creativity, safety and health professionals should "set the table" with consistent and positive reinforcement encouraging employee participation without fear of reprisal or embarrassment. Employees should be encouraged to bring forward their ideas and concepts with the assurance that their idea or concept would be provided appropriate consideration and evaluation. This positive reinforcement should encourage each and every employee to accept responsibility for his or her segment of the safety and health function while actively working to improve the safety and health function as a whole.

ENERGIZING YOUR EMPLOYEE'S CREATIVITY IN SAFETY AND HEALTH

1. Safety and health is *our* program. Remove the word *my* from your vocabulary.
2. *Always* provide timely feedback for any and all ideas, questions, etc.

---

[*] "A Theory of Human Motivation," A.H. Maslow, *Psychology Review*, 50(4), 370–396, 1943. Also see http://psychclassics.yorku.ca/Maslow/motivation.htm. In general, Maslow identified the hierarchy beginning with physiological and moving upward within the hierarchy to safety, love/belonging, esteem, and ending with self-actualization.

[†] Id.

[‡] Id.

3. *Use* your employee's ideas. Who knows the job better than the employee who works at the job each day.
4. *Acknowledge* employees for their ideas.
5. *Give credit* to employees for their ideas.
6. *Promote* active involvement in safety and health.
7. *Share* appropriate safety and health data with employees.
8. *Continuously* educate employees at their individual educational level.
9. *Actively involve* employees in appropriate safety and health decision making.
10. *Create* and maintain transparency within the safety and health function.
11. *Never* embarrass an employee, especially in front of his or her peers.
12. *Praise* employees for outstanding efforts.
13. A pat on the back is worth more than a dollar in the paycheck.
14. *Always* be positive.
15. *Trust*, once lost, is seldom regained.

Through active involvement and empowerment of employees within the safety and health function, employees will soon achieve a sense of ownership in their safety and health as well as the safety and health program for the organization. This "cultural shift" in your employee's attitude and way of thinking about safety and health is exactly what safety and health professionals should strive to achieve in order to truly create a top-notch safety and health program. Upon this acceptance of ownership of the safety and health function by your employees, safety and health will become a priority and your employees will begin to view safety and health from the ownership prospective. At this point in the process, your employees will begin to develop a critical and creative prospective, and the ideas and concepts for improvement within the safety and health function will emerge.

Let's take a relatively simple process such as selecting employees' personal protective equipment. After the safety professional analyzes the job functions and identifies the type and requirements for necessary personal protective equipment and other essential requirements, he or she has a great opportunity to acquire active involvement and stimulate employee creativity through involvement of employees in the PPE selection process. The safety professional can acquire a number of different types of PPE that meet the regulatory requirements and exhibit this equipment for employee's thoughts, ideas, and opinions. Employees can be encouraged to become actively involved by providing the PPE for their evaluation and testing within the various job functions. Each employee should be provided an opportunity in writing or orally to present his or her opinion of the comfort, workability, and other factors within the job function and provide his or her thoughts and ideas regarding the PPE. The process of permitting employees to touch, feel, test, and express their opinion permits employees to buy in to the process as well as serves as an educational opportunity for employees to learn the importance of the PPE in the workplace.

Safety professionals should look for opportunities in their everyday activities through which to actively involve their employees and stimulate their creative juices. Who knows the specific job better than the employee who performs the function daily? It is our job to unleash the creative juices and expertise housed within each employee and utilize this knowledge and creativity to enhance the safety and health function. Each and every employee is a storehouse of information; however, it is our jobs, as safety and health professionals, to provide the opportunities for our employees, positively reinforce their efforts, and utilize their ideas to create and maintain a safe and healthy work environment.

DISCUSSION QUESTIONS

1. Explain Maslow's hierarchy of needs as it applies to the safety and health function.
2. Identify and discuss at least one method through which to energize and empower employees within the safety and health function.
3. Explain how a safety and health professional can acquire buy-in by employees.
4. Please provide one example of how a safety and health professional can energize employees' creativity within the safety and health function.

9 How May I Help You?

Management is the art of getting three men to do three men's work.

—**William Feathers**

When a person is down in the world, an ounce of help is better than a pound of preaching.

—**Edward Bulwer-Lytton**

LEARNING OBJECTIVES

1. Assess and analyze the impact of other laws and regulations on the safety and health function.
2. Assess how the phrase "How can I help you?" opens the doors to effective communication with employees.
3. Analyze potential legal and ethical implications involving employee communications.

The safety and health function does not work in a vacuum. Virtually every issue that emerges with most operations possesses some aspect that touches the safety and health function. Additionally, many human resource, production, and quality issues can be "twisted" into a safety and health issue. Safety and health professionals should be aware of the numerous correlating laws and regulations that can impact the safety and health function and remember to start any review of a pending issue from a very broad prospective and narrow the scope and viability of the issue throughout the review process. Each and every potential law or regulation that may impact the particular issue should be examined for applicability to the specific situation and either rejected or included for further examination prior to a final decision. "Shooting from the hip" is a sure-fire way through which a safety and health professional can make a wrong decision and get him or her into hot water.

When an employee confronts a safety and health professional with an issue, one easy way in which a safety and health professional can open the lines of communication, defuse any potential hostilities, and ensure that all of the facts are acquired is simply to ask, "How can I help you?" and *listen* to the employee or individual. This simple question can open the doors to information that can be vital within your decision-making process. And safety and health professionals should listen intently to each and every word provided by the employee or individual while identifying

73

the various legal or regulatory issues that are identified within the conversation. For example:

> An employee stops you on the production floor while you are conducting an inspection. The employee asks to talk to you about a safety issue impacting production. You respond, "How can I help you?" The employee communicates that production has been slowed due to a guard being removed from a machine. The employee further states that she has told her supervisor several times, and he only responds with sexist remarks. She wanted to tell you because this situation and her reporting this to her supervisor has resulted in her being transferred to another job and her overtime cut, which affects her pay and childcare situation. As you listen intently, she informs you that this is a safety issue that should be addressed in order that someone else is not seriously injured like she was on the machine last year.

Can you identify the myriad laws and regulations that could be applicable and possibly impact the company or organization in the example above? How about Occupational Safety and Health Administration (OSHA) regulations, the Americans with Disabilities Act (ADA), the Family and Medical Leave Act (FMLA), Wage and Hour, and Title VII of the Civil Rights Act, among other potential issues. With the safety and health professional listening intently after opening the conversation with "How can I help you?" the safety and health professional potentially could identify a number of issues that could evolve to be injurious not only to the safety and health function but also to the company or organization. Early identification and detection can often avoid costly claims and litigation in the future. This simple question, "How can I help you?" opened the door and permitted the actively listening safety and health professional to identify potential safety issues as well as other related issues that could impact the company or organization directly or indirectly.

When a safety and health professional opens the door with "How can I help you?" employees often provide information in a rapid-fire format and include information that may or may not be pertinent for other functions within the organization. It is essential that the safety and health professional listen intently to the employee and, if necessary, return and summarize the information for the employee or ask the employee to repeat the information to ensure complete clarity. What is the safety and health professional's duty and responsibilities if and when reporting such information within the organization or company? Outside of the company or organization?

As Paul Harvey used to say in his broadcasts, safety and health professionals should be prepared for "the rest of the story" when talking with employees. After opening the door with the statement "How can I help you?" safety and health professionals should be prepared to for the employee to start with a safety issue or complaint. The employee is often attempting to summarize the issue or provide his or her opinion or interpretation of the particular safety issue. Prudent safety and health professionals should listen intently and ask questions, but refrain from providing an opinion or rendering a decision. There is virtually always more to the issue or story.

The information provided by employees can often place the safety and health professional in a dilemma as to whether to report or not to report. On one hand, if the safety and health professional runs to management with each and every potential issue proclaimed by employees, he or she can lose credibility with management. Additionally, the employee may have come to the safety and health professional assuming the conversation would be in confidence. The sharing of this information may cause the employee to lose trust in the safety and health professional. Conversely, if the safety and health professional is informed of an issue that potentially could impact the operation, management, regulatory compliance, or future litigation against the organization, does the safety and health professional owe a duty to his or her organization to report the information? This is often a judgment call that must be made by the safety and health professional depending on many variables, including credibility, circumstances, and other factors. The most prudent route for many safety and health professionals is to report the information to the appropriate department or manager as proscribed within the company or organization's policies and procedures.

With any identification of a safety and health issue by an employee, it is essential that the safety and health professional follow up and provide a response to the employee(s). Failure to investigate and provide a response, whether positive or negative, to the employee sends a signal that the employee's issue was not important or the safety and health professional does not care about the issue. Safety and health professionals should not fear providing a negative response to the employee regarding the issue. The simple act of providing timely feedback to the employee with an adequate explanation of the reasons for the decision is often adequate. Additionally, this simple act of proving timely feedback to the employee shows professionalism and a true concern for your employees.

Safety and health professionals should be cognizant of their status within the management team and organizational structure and refrain from shooting from the hip with regards to any decision-making issue. This is especially important when the potential of disciplinary action is involved in the situation. The progressive disciplinary policy and procedure in many organizations provides disciplinary authority to the safety and health professional, while in other organizations the safety and health professional possesses no disciplinary authority. It is important that the safety and health professional know the limits of his or her authority within the organization and follow the policy and procedure to the letter. Decisions, especially when the potential of disciplinary action is involved, should never be made in haste and "in the heat of the battle." Safety and health professionals should completely and thoroughly investigate each and every situation where disciplinary action may be involved and be absolutely correct before issuing any level of disciplinary action within the safety and health area.

When the door is opened with "How can I help you?" safety and health professionals should be prepared for the potential of learning far more than they would like to know, which can create ethical as well as professional issues. Safety and health professionals should identify the appropriate avenues within their organization through which to transfer information while protecting the employee and themselves. When employees trust the safety and health professional, information that far exceeds the safety and health area can be provided by employees. Knowing when and where to provide this information as well as protecting the source and himself or herself is a mark of professionalism for the safety and health professional.

DISCUSSION QUESTIONS

1. Are there any other phrases, such as "How can I help you? that safety and health professionals can utilize to open conversations with employees?
2. Identify one law or regulation that could impact the safety and health function, and explain how the law or regulation impacts the function.
3. Provide one example where a safety and health professional could encounter an ethical issue in the performance of his or her job function.

10 Rethinking the Employment Relationship

Good management consists in showing average people how to do the work of superior people.

—John D. Rockefeller

Man is the principal syllable in management.

—C.T. McKenzie

LEARNING OBJECTIVES

1. Identify the various employment statuses that can be on a worksite.
2. Assess and analyze the various employment status categories.
3. Analyze and assess the OSHA regulations for multi-employer workplaces.

When a safety and health professional walks onto a job site, does the safety and health professional know the employment status of each and every individual working on the site? Does this make a difference from a safety and health perspective? Does it make a difference from a legal perspective? Does it make a difference from a management perspective? Let's examine a typical construction site. Your company or organization may have employees performing work at the site. Contractors will be working at the site with subcontractors hired by the general contractor to perform specific work. Your company or the contractor may have individuals hired through a temporary employment agency to perform specific work activities. All of these individuals are performing various work functions at the worksite. What is the distinction among and between each of these individuals working at the worksite, and how does it impact the safety and health professional is his or her job function?

From a legal perspective, safety and health professionals should be aware that there is a significant difference among and between the legal statuses of these individuals working at the job site. Independent contractors, as the name implies, work independently within a contractual relationship and contract, usually written, to perform specific work activities. Contractors are independent and usually paid by the job or activity. Contractors are responsible for their own workers' compensation, insurance, wages, taxes, and other federal, state, and local requirements. Safety and health compliance at the worksite can be a provision within the written contact document. Safety and health professionals usually cannot discipline for failure to

77

comply with safety and health-related company policies but would be required to return to the provisions of the written contract or pursue a breach of the contract. Subcontractors work directly for the contractor under a written or oral agreement. Subcontractors, like contractors, are usually independent and usually paid by the job. Subcontractors, again as the name implies, are usually smaller or more specialized contractors who work under a written or oral agreement or contract with the general contractor. As an example, an independent plumber may be a subcontractor to a home builder for the purpose of installing the kitchen within the house.

The use of temporary workers, or "temps," is often utilized by companies or organizations for short-term work. Companies or organizations can contract with an employment agency or other entity to provide qualified labor for a specified time period for usually a fixed price. The workers may be working at your company or organization but are the employees of the employment agency or other entity. The employment agency is responsible for wages, hiring and termination, workers' compensation, insurance, taxes and other federal, state, and local requirements. Safety and health training and compliance at the worksite are often provisions within the written agreement between the contracting company or organization and the employment agency. Some agreements provide that temporary workers may transition to and be hired as employees of the company or organization after a period of satisfactory work performance.

According to *Black's Law Dictionary*, an employee is "a person in the service of another under any contract of hire, express or implied, oral or written, where the employer has the power or right to control and direct the employee in the material details of how the work is to be performed."* An employee has usually been hired by your company to perform ongoing work activities; however, there are different types of legal perimeters surrounding different types of employees within the company or organization. Probationary employees are often newly hired individuals who must successfully complete a 30- to 90-day or longer probationary period before becoming a full-time employee. Companies and organizations can also employ part-time employees who work a limited number of hours and are usually provided no benefits, seasonal employees who work only for a period of time each year, and interns who are often students working for a specified period of time while learning.

Within the employee ranks of many companies or organizations, the different levels are often delineated by whether the employee is paid on an hourly basis (nonexempt) or salary basis (exempt). Most managerial employees are paid on a salary basis and are considered within the executive, administrative, or professional exemptions of the Fair Labor Standards Act.† Employees paid on an hourly basis are usually encompassed within the Fair Labor Standards Act, which also addresses minimum wage rates, overtime provisions, and other requirements.‡

Whether exempt or nonexempt employees, most employees are considered "at will" employees, which means that an employee can quit at any time; however, the employer may terminate the employee at any time and for any reason. There are state specific

---

* *Black's Law Dictionary*, 5th edition (1983).
† 29 USC 201, *et. seq.* (1938).
‡ Id.

exceptions to the at-will employment doctrine; however, the vast majority of employees, both salaries and hourly, work under the at-will employment doctrine. However, employees working under a collective bargaining agreement (commonly referred to as a union contract) are not under the at-will employment doctrine and must abide by the terms and conditions of the collective bargaining agreement or contract.

In general, the above are the various employment statuses that can be working on any given worksite at the same time with little or no distinction between the various them. The specific employment categorization can become important for safety and health professionals in the management of the safety function, the issuance of disciplinary action for safety and health violations, in the event of an accident, during an Occupational Safety and Health Administration (OSHA) inspection, and many other circumstances. The critical question for many safety and health professionals is how to successfully manage the safety and health function in situations where there can be many different employers and many different statuses with each possessing different requirements and regulations.

As identified below, the Occupational Safety and Health Administration has addressed the issue of many different employers and employees with different employment statuses through Directive Number CPL 02-00-124 titled "Multi-Employer Citation Policy," issued in 1999. Through this directive, the compliance officer is provided a methodology through which to identify and issue citations to one or all employers working on one worksite for identifiable safety and health violations. To address this potential risk, prudent safety and health professionals adopted the preparatory tact to either manage all employees, contractors, and others working on the site requiring the same level of safety and health compliance as within the operations or adopt a hands-off approach for on-site contractors and subcontractors with the risk and responsibility for safety and health, as well as citations being shifted to the contractor or subcontractor through the written contract language. The question for safety and health professionals when they view a workplace under their jurisdiction is, why are we making this type of distinction in employment status, and is it in the best interest of the individual worker to address safety and health from a bifurcated management perspective? From an OSHA and workers' compensation perspective, the company or organization would only be responsible for "employees." Contractors and subcontractors would be under contract and thus responsible for their employees under OSHA and individual state workers' compensation laws. From a financial perspective, it may be beneficial to use contractors and subcontractors for the specific project. From an expertise or certification basis, it may be beneficial to contract for the specific expertise rather than hiring an employee. However, from an efficacy perspective, if a contractor or subcontractor incurs a severe injury or OSHA cites the workplace, it will often be the company or organization's name in the newspaper or on television.

In 2010, the Patient Protection and Affordable Care Act (PPACA)* (commonly called Obamacare or the Affordable Care Act (ACA)) was signed into law. This new law, in general, is a major change in the U.S. health care system addressing affordability of health insurance, preexisting conditions, and a number of major

---

* Pub. L. 111-148 (March 23, 2010).

changes to permit health insurance coverage for virtually every American. Under the current health care system, companies or organizations offer coverage for employees, and others, such as subcontractors working independently, potentially could purchase coverage directly from the insurance carriers. Many workers, often part-time or even full-time, do not currently have health insurance but should be eligible in the future. For safety and health professionals, the Patient Protection and Affordable Care Act possesses the potential of creating a new dynamic that cannot be overlooked. How will the Patient Protection and Affordable Care Act impact the current state-operated workers' compensation laws and regulations? If every employee is provided health care coverage, will the current workers' compensation system morph into a Canadian-type universal care system? Conceptually, will there continue to be a need for employers to provide workers' compensation coverage for their employees if every employee is covered under the PPACA?

The cost of workers' compensation for employees has steadily risen over the past decades and has often been a driver for companies and organizations to invest in safety and health professionals, safety equipment, safety and health training, etc. If the current driver of workers' compensation costs is removed from employers as a result of the PPACA, will the safety and health of the employees and the workplace receive a reduced priority by management? Will employees be more prone to take greater risks in their job knowing that they possess health insurance? Will companies and organizations be more willing to accept greater risks within their operations or increase production rates if the correlating costs of workers' compensation is removed from their balance sheet?

At this point in time, the only thing certain is that the health care system is going to change. Prudent safety and health professionals should not be complacent is identifying any changes and any potential impacts on the safety and health functions and related functions. The PPACA, although currently controversial, possesses the potential of having an impact, whether positive or negative, on the safety and health profession.

Record Type: Instruction
  - Directive Number: CPL 02-00-124
  - Old Directive Number: CPL 2-0.124
  - Title: Multi-Employer Citation Policy.
  - Information Date: 12/10/1999

U.S. DEPARTMENT OF LABOR   **Occupational Safety and Health Administration**

DIRECTIVE NUMBER:CPL 2-0.124	**EFFECTIVE DATE:** December 10, 1999
SUBJECT: Multi-Employer Citation Policy	

ABSTRACT

Purpose: To Clarify the Agency's multi-employer citation policy

Scope: OSHA-wide

References: OSHA Instruction CPL 2.103 (the FIRM)

Suspensions: Chapter III, Paragraph C. 6. of the FIRM is suspended and
                          replaced by this directive

State Impact: This Instruction describes a Federal Program Change.
                          Notification of State intent is required, but adoption is not.

Action Offices: National, Regional, and Area Offices

Originating Office: Directorate of Compliance Programs

Contact: Carl Sall (202) 693-2345
                          Directorate of Construction
                          N3468 FPB
                          200 Constitution Ave., NW
                          Washington, DC 20210

By and Under the Authority of
R. Davis Layne
Deputy Assistant Secretary, OSHA

TABLE OF CONTENTS

I.   Purpose. This Directive clarifies the Agency's multi-employer citation policy and suspends Chapter III. C. 6. of OSHA's Field Inspection Reference Manual (FIRM).

II.   Scope. OSHA-Wide

III.   Suspension. Chapter III. Paragraph C. 6. of the FIRM (CPL 2.103) is suspended and replaced by this Directive.

IV.   References. OSHA Instructions:
       ○   CPL 02-00.103; OSHA Field Inspection Reference Manual (FIRM), September 26, 1994.
       ○   ADM 08-0.1C, OSHA Electronic Directive System, December 19,1997.

V.   Action Information

A. <u>Responsible Office.</u> Directorate of Construction.
B. <u>Action Offices.</u> National, Regional and Area Offices
C. <u>Information Offices.</u> State Plan Offices, Consultation Project Offices

<u>Federal Program Change.</u> This Directive describes a Federal Program Change for which State adoption is not required. However, the States shall respond via the two-way memorandum to the Regional Office as soon as the State's intent regarding the multi-employer citation policy is known, but no later than 60 calendar days after the date of transmittal from the Directorate of Federal-State Operations.

<u>Force and Effect of Revised Policy.</u> The revised policy provided in this Directive is in full force and effect from the date of its issuance. It is an official Agency policy to be implemented OSHA-wide.

<u>Changes in Web Version of FIRM</u> A note will be included at appropriate places in the FIRM as it appears on the Web indicating the suspension of Chapter III paragraph 6. C. and its replacement by this Directive, and a hypertext link will be provided connecting viewers with this Directive.

<u>Background.</u> OSHA's Field Inspection Reference Manual (FIRM) of September 26, 1994 (CPL 2.103), states at Chapter III, paragraph 6. C., the Agency's citation policy for multi-employer worksites. The Agency has determined that this policy needs clarification. This directive describes the revised policy.

- <u>Continuation of Basic Policy.</u> This revision continues OSHA's existing policy for issuing citations on multi-employer worksites. However, it gives clearer and more detailed guidance than did the earlier description of the policy in the FIRM, including new examples explaining when citations should and should not be issued to exposing, creating, correcting, and controlling employers. These examples, which address common situations and provide general policy guidance, are not intended to be exclusive. In all cases, the decision on whether to issue citations should be based on all of the relevant facts revealed by the inspection or investigation.

A. <u>No Changes in Employer Duties.</u> This revision neither imposes new duties on employers nor detracts from their existing duties under the OSH Act. Those duties continue to arise from the employers' statutory duty to comply with OSHA standards and their duty to exercise reasonable diligence to determine whether violations of those standards exist.

<u>Multi-employer Worksite Policy.</u> The following is the multi-employer citation policy:

- <u>Multi-employer Worksites.</u> On multi-employer worksites (in all industry sectors), more than one employer may be citable for a hazardous condition that violates an OSHA standard. A two-step process must be followed in determining whether more than one employer is to be cited.

1. <u>Step One.</u> The first step is to determine whether the employer is a creating, exposing, correcting, or controlling employer. The definitions in paragraphs (B) - (E) below explain and give examples of each. Remember that an employer may have multiple roles (see paragraph H). Once you determine the role of the employer, go to Step Two to determine if a citation is appropriate (NOTE: only

exposing employers can be cited for General Duty Clause violations).

2. <u>Step Two</u>. If the employer falls into one of these categories, it has obligations with respect to OSHA requirements. Step Two is to determine if the employer's actions were sufficient to meet those obligations. The extent of the actions required of employers varies based on which category applies. Note that the extent of the measures that a controlling employer must take to satisfy its duty to exercise reasonable care to prevent and detect violations is less than what is required of an employer with respect to protecting its own employees.

A. <u>The Creating Employer</u>

1. <u>Step 1</u>: Definition: The employer that caused a hazardous condition that violates an OSHA standard.

2. <u>Step 2</u>: Actions Taken: Employers must not create violative conditions. An employer that does so is citable even if the only employees exposed are those of other employers at the site.

   a. **Example 1:** Employer Host operates a factory. It contracts with Company S to service machinery. Host fails to cover drums of a chemical despite S's repeated requests that it do so. This results in airborne levels of the chemical that exceed the Permissible Exposure Limit.

      Analysis: **Step 1:** Host is a creating employer because it caused employees of S to be exposed to the air contaminant above the PEL. **Step 2:** Host failed to implement measures to prevent the accumulation of the air contaminant. It could have met its OSHA obligation by implementing the simple engineering control of covering the drums. Having failed to implement a feasible engineering control to meet the PEL, Host is citable for the hazard.

   b. **Example 2:** Employer M hoists materials onto Floor 8, damaging perimeter guardrails. Neither its own employees nor employees of other employers are exposed to the hazard. It takes effective steps to keep all employees, including those of other employers, away from the unprotected edge and informs the controlling employer of the problem. Employer M lacks authority to fix the guardrails itself.

      Analysis: **Step 1:** Employer M is a creating employer because it caused a hazardous condition by damaging the guardrails. **Step 2:** While it lacked the authority to fix the guardrails, it took immediate and effective steps to keep all employees away from the hazard and notified the controlling employer of the hazard. Employer M is not citable since it took effective measures to prevent employee exposure to the fall hazard.

B. <u>The Exposing Employer</u>
   1. <u>Step 1: Definition</u>: An employer whose own employees are exposed to the hazard. See Chapter III, section (C)(1)(b) for a discussion of what constitutes exposure.
   2. <u>Step 2: Actions taken</u>: If the exposing employer created the violation, it is citable for the violation as a creating employer. If the violation was created by another employer, the exposing employer is citable if it (1) knew of the hazardous condition or failed to exercise reasonable diligence to discover the condition, and (2) failed to take steps consistent with its authority to protect is employees. If the exposing employer has authority to correct the hazard, it must do so. If the exposing employer lacks the authority to correct the hazard, it is citable if it fails to do each of the following: (1) ask the creating and/or controlling employer to correct the hazard; (2) inform its employees of the hazard; and (3) take reasonable alternative protective measures. In extreme circumstances (e.g., imminent danger situations), the exposing employer is citable for failing to remove its employees from the job to avoid the hazard.
      a. **Example 3:** Employer Sub S is responsible for inspecting and cleaning a work area in Plant P around a large, permanent hole at the end of each day. An OSHA standard requires guardrails. There are no guardrails around the hole and Sub S employees do not use personal fall protection, although it would be feasible to do so. Sub S has no authority to install guardrails. However, it did ask Employer P, which operates the plant, to install them. P refused to install guardrails.

         Analysis: Step 1: Sub S is an exposing employer because its employees are exposed to the fall hazard. **Step 2:** While Sub S has no authority to install guardrails, it is required to comply with OSHA requirements to the extent feasible. It must take steps to protect its employees and ask the employer that controls the hazard - Employer P - to correct it. Although Sub S asked for guardrails, since the hazard was not corrected, Sub S was responsible for taking reasonable alternative protective steps, such as providing personal fall protection. Because that was not done, Sub S is citable for the violation.

      b. **Example 4:** Unprotected rebar on either side of an access ramp presents an impalement hazard. Sub E, an electrical subcontractor, does not have the authority to cover the rebar. However, several times Sub E asked the general contractor, Employer GC, to cover the rebar. In the meantime, Sub E instructed its employees to use a different access route that avoided most of the uncovered rebar and required them to keep as far from the rebar as possible.

Analysis: Step 1: Since Sub E employees were still exposed to some unprotected rebar, Sub E is an exposing employer. **Step 2:** Sub E made a good faith effort to get the general contractor to correct the hazard and took feasible measures within its control to protect its employees. Sub E is not citable for the rebar hazard.

C. The Correcting Employer
   1. Step 1: Definition: An employer who is engaged in a common undertaking, on the same worksite, as the exposing employer and is responsible for correcting a hazard. This usually occurs where an employer is given the responsibility of installing and/or maintaining particular safety/health equipment or devices.
   2. Step 2: Actions taken: The correcting employer must exercise reasonable care in preventing and discovering violations and meet its obligations of correcting the hazard.
      a. **Example 5**: Employer C, a carpentry contractor, is hired to erect and maintain guardrails throughout a large, 15-story project. Work is proceeding on all floors. C inspects all floors in the morning and again in the afternoon each day. It also inspects areas where material is delivered to the perimeter once the material vendor is finished delivering material to that area. Other subcontractors are required to report damaged/missing guardrails to the general contractor, who forwards those reports to C. C repairs damaged guardrails immediately after finding them and immediately after they are reported. On this project few instances of damaged guardrails have occurred other than where material has been delivered. Shortly after the afternoon inspection of Floor 6, workers moving equipment accidentally damage a guardrail in one area. No one tells C of the damage and C has not seen it. An OSHA inspection occurs at the beginning of the next day, prior to the morning inspection of Floor 6. None of C's own employees are exposed to the hazard, but other employees are exposed.

      Analysis: Step 1: C is a correcting employer since it is responsible for erecting and maintaining fall protection equipment. **Step 2:** The steps C implemented to discover and correct damaged guardrails were reasonable in light of the amount of activity and size of the project. It exercised reasonable care in preventing and discovering violations; it is not citable for the damaged guardrail since it could not reasonably have known of the violation.

D. The Controlling Employer

1. Step 1: Definition: An employer who has general supervisory authority over the worksite, including the power to correct safety and health violations itself or require others to correct them. Control can be established by contract or, in the absence of explicit contractual provisions, by the exercise of control in practice. Descriptions and examples of different kinds of controlling employers are given below.

2. Step 2: Actions Taken: A controlling employer must exercise reasonable care to prevent and detect violations on the site. The extent of the measures that a controlling employer must implement to satisfy this duty of reasonable care is less than what is required of an employer with respect to protecting its own employees. This means that the controlling employer is not normally required to inspect for hazards as frequently or to have the same level of knowledge of the applicable standards or of trade expertise as the employer it has hired.

3. Factors Relating to Reasonable Care Standard. Factors that affect how frequently and closely a controlling employer must inspect to meet its standard of reasonable care include:

    a. The scale of the project;

    b. The nature and pace of the work, including the frequency with which the number or types of hazards change as the work progresses;

    c. How much the controlling employer knows both about the safety history and safety practices of the employer it controls and about that employer's level of expertise.

    d. More frequent inspections are normally needed if the controlling employer knows that the other employer has a history of non-compliance. Greater inspection frequency may also be needed, especially at the beginning of the project, if the controlling employer had never before worked with this other employer and does not know its compliance history.

    e. Less frequent inspections may be appropriate where the controlling employer sees strong indications that the other employer has implemented effective safety and health efforts. The most important indicator of an effective safety and health effort by the other employer is a consistently high level of compliance. Other indicators include the use of an effective, graduated system of enforcement for non-compliance with safety and health requirements coupled with regular jobsite safety meetings and safety training.

4. Evaluating Reasonable Care. In evaluating whether a controlling employer has exercised reasonable care in preventing and discovering violations, consider questions such as whether the controlling employer:

    a. Conducted periodic inspections of appropriate frequency (frequency should be based on the

factors listed in G.3.);
   b.   Implemented an effective system for promptly correcting hazards;
   c.   Enforces the other employer's compliance with safety and health require-
      ments with an effective, graduated system of enforcement and follow-up
      inspections.

5.   Types of Controlling Employers
   a.   Control Established by Contract. In this case, **the Employer Has a
      Specific Contract Right to Control Safety:** To be a controlling employer,
      the employer must itself be able to prevent or correct a violation or to
      require another employer to prevent or correct the violation. One source
      of this ability is explicit contract authority. This can take the form of a
      specific contract right to require another employer to adhere to safety and
      health requirements and to correct violations the controlling employer
      discovers.

      (1) **Example 6:** Employer GH contracts with Employer S to do sand-
          blasting at GH's plant. Some of the work is regularly scheduled
          maintenance and so is general industry work; other parts of the proj-
          ect involve new work and are considered construction. Respiratory
          protection is required. Further, the contract explicitly requires S to
          comply with safety and health requirements. Under the contract GH
          has the right to take various actions against S for failing to meet con-
          tract requirements, including the right to have non-compliance cor-
          rected by using other workers and back-charging for that work. S is
          one of two employers under contract with GH at the work site, where
          a total of five employees work. All work is done within an existing
          building. The number and types of hazards involved in S's work do
          not significantly change as the work progresses. Further, GH has
          worked with S over the course of several years. S provides periodic
          and other safety and health training and uses a graduated system of
          enforcement of safety and health rules. S has consistently had a high
          level of compliance at its previous jobs and at this site. GH monitors
          S by a combination of weekly inspections, telephone discussions and
          a weekly review of S's own inspection reports. GH has a system of
          graduated enforcement that it has applied to S for the few safety and
          health violations that had been committed by S in the past few years.
          Further, due to respirator equipment problems S violates respira-
          tory protection requirements two days before GH's next scheduled
          inspection of S. The next day there is an OSHA inspection. There is
          no notation of the equipment problems in S's inspection reports to
          GH and S made no mention of it in its telephone discussions.

Analysis: Step 1: GH is a controlling employer because it has general supervisory authority over the worksite, including contractual authority to correct safety and health violations. **Step 2**: GH has taken reasonable steps to try to make sure that S meets safety and health requirements. Its inspection frequency is appropriate in light of the low number of workers at the site, lack of significant changes in the nature of the work and types of hazards involved, GH's knowledge of S's history of compliance and its effective safety and health efforts on this job. GH has exercised reasonable care and is not citable for this condition.

(2)   **Example 7**: Employer GC contracts with Employer P to do painting work. GC has the same contract authority over P as Employer GH had in Example 6. GC has never before worked with P. GC conducts inspections that are sufficiently frequent in light of the factors listed above in (G)(3). Further, during a number of its inspections, GC finds that P has violated fall protection requirements. It points the violations out to P during each inspection but takes no further actions.

Analysis: Step 1: GC is a controlling employer since it has general supervisory authority over the site, including a contractual right of control over P. **Step 2**: GC took adequate steps to meet its obligation to discover violations. However, it failed to take reasonable steps to require P to correct hazards since it lacked a graduated system of enforcement. A citation to GC for the fall protection violations is appropriate.

(3)   **Example 8**: Employer GC contracts with Sub E, an electrical subcontractor. GC has full contract authority over Sub E, as in Example 6. Sub E installs an electric panel box exposed to the weather and implements an assured equipment grounding conductor program, as required under the contract. It fails to connect a grounding wire inside the box to one of the outlets. This incomplete ground is not apparent from a visual inspection. Further, GC inspects the site with a frequency appropriate for the site in light of the factors discussed above in (G)(3). It saw the panel box but did not test the outlets to determine if they were all grounded because Sub E represents that it is doing all of the required tests on all receptacles. GC knows that Sub E has implemented an effective safety and health program. From previous experience it also knows Sub E is familiar with the applicable safety requirements and is technically competent. GC had asked Sub E if the electrical equipment is OK for use and was assured that it is.

Analysis: Step 1: GC is a controlling employer since it has general supervisory authority over the site, including a contractual right of control over Sub E. **Step 2:** GC exercised reasonable care. It had determined that Sub E had technical expertise, safety knowledge and had implemented safe work practices. It conducted inspections with appropriate frequency. It also made some basic inquiries into the safety of the electrical equipment. Under these circumstances GC was not obligated to test the outlets itself to determine if they were all grounded. It is not citable for the grounding violation.

b.   <u>Control Established by a Combination of Other Contract Rights</u>: Where there is no explicit contract provision granting the right to control safety, or where the contract says the employer does <u>not</u> have such a right, an employer may still be a controlling employer. The ability of an employer to control safety in this circumstance can result from a combination of contractual rights that, together, give it broad responsibility at the site involving almost all aspects of the job. Its responsibility is broad enough so that its contractual authority necessarily involves safety. The authority to resolve disputes between subcontractors, set schedules and determine construction sequencing are particularly significant because they are likely to affect safety. (NOTE: citations should only be issued in this type of case after consulting with the Regional Solicitor's office).

(1)   **Example 9:** Construction manager M is contractually obligated to: set schedules and construction sequencing, require subcontractors to meet contract specifications, negotiate with trades, resolve disputes between subcontractors, direct work and make purchasing decisions, which affect safety. However, the contract states that M does not have a right to require compliance with safety and health requirements. Further, Subcontractor S asks M to alter the schedule so that S would not have to start work until Subcontractor G has completed installing guardrails. M is contractually responsible for deciding whether to approve S's request.

Analysis: Step 1: Even though its contract states that M does not have authority over safety, the combination of rights actually given in the contract provides broad responsibility over the site and results in the ability of M to direct actions that necessarily affect safety. For example, M's contractual obligation to determine whether to approve S's request to alter the schedule has direct safety implications. M's decision relates directly to whether S's employees will be protected from a fall hazard. M is a controlling employer. **Step 2:** In this example, if M refused to alter the schedule, it would be citable for the fall hazard violation.

(2)   **Example 10:** Employer ML's contractual authority is limited to reporting on subcontractors' contract compliance to owner/developer O and making contract payments. Although it reports on the extent to which the subcontractors are complying with safety and health infractions to O, ML does not exercise any control over safety at the site.

Analysis: Step 1: ML is not a controlling employer because these contractual rights are insufficient to confer control over the subcontractors and ML did not exercise control over safety. Reporting safety and health infractions to another entity does not, by itself (or in combination with these very limited contract rights), constitute an exercise of control over safety. **Step 2:** Since it is not a controlling employer it had no duty under the OSH Act to exercise reasonable care with respect to enforcing the subcontractors' compliance with safety; there is therefore no need to go to Step 2.

c.   <u>Architects and Engineers</u>: Architects, engineers, and other entities are controlling employers only if the breadth of their involvement in a construction project is sufficient to bring them within the parameters discussed above.

(1)   **Example 11:** Architect A contracts with owner O to prepare contract drawings and specifications, inspect the work, report to O on contract compliance, and to certify completion of work. A has no authority or means to enforce compliance, no authority to approve/reject work and does not exercise any other authority at the site, although it does call the general contractor's attention to observed hazards noted during its inspections.

Analysis: Step 1: A's responsibilities are very limited in light of the numerous other administrative responsibilities necessary to complete the project. It is little more than a supplier of architectural services and conduit of information to O. Its responsibilities are insufficient to confer control over the subcontractors and it did not exercise control over safety. The responsibilities it does have are insufficient to make it a controlling employer. Merely pointing out safety violations did not make it a controlling employer. NOTE: In a circumstance such as this it is likely that broad control over the project rests with another entity. **Step 2:** Since A is not a controlling employer it had no duty under the OSH Act to exercise reasonable care with respect to enforcing the subcontractors' compliance with safety; there is therefore no need to go to Step 2.

(2)   **Example 12:** Engineering firm E has the same contract authority and functions as in Example 9.

Analysis: Step 1: Under the facts in Example 9, E would be considered a controlling employer. **Step 2:** The same type of analysis described in Example 9 for Step 2 would apply here to determine if E should be cited.

d.   Control Without Explicit Contractual Authority. Even where an employer has no explicit contract rights with respect to safety, an employer can still be a controlling employer if, in actual practice, it exercises broad control over subcontractors at the site (see Example 9). NOTE: Citations should only be issued in this type of case after consulting with the Regional Solicitor's office.

(1)   **Example 13:** Construction manager MM does not have explicit contractual authority to require subcontractors to comply with safety requirements, nor does it explicitly have broad contractual authority at the site. However, it exercises control over most aspects of the subcontractors' work anyway, including aspects that relate to safety.

Analysis: Step 1: MM would be considered a controlling employer since it exercises control over most aspects of the subcontractor's work, including safety aspects. **Step 2:** The same type of analysis on reasonable care described in the examples in (G)(5)(a) would apply to determine if a citation should be issued to this type of controlling employer.

E.   Multiple Roles

1.   A creating, correcting or controlling employer will often also be an exposing employer. Consider whether the employer is an exposing employer before evaluating its status with respect to these other roles.

2.   Exposing, creating and controlling employers can also be correcting employers if they are authorized to correct the hazard.

---

* OSHA website: www.OSHA.gov.

DISCUSSION QUESTIONS

1. Please apply the OSHA multiemployer worksite rule to a hypothetical construction worksite. Please explain your answer in detail.
2. Please identify the various employment status categories and compare and contrast each category.
3. What is the difference between a probationary employee and a full-time employee? A contractor?
4. What potential impacts can the Affordable Care Act have upon the safety and health function?

11 Safety Impacts Everything

When prosperity comes, do not use all of it.

—Confucius

The man without a purpose is like a ship without a rudder—waif, a nothing, a no man.

—Thomas Carlyle

LEARNING OBJECTIVES

1. Analyze and identify the various functions impacted by safety and health.
2. Identify and assess the impact of other laws on the safety and health function.
3. Analyze and assess the laws and regulations enforced by the Equal Employment Opportunity Commission (EEOC).
4. Analyze and assess the laws and regulations enforced by the U.S. Department of Labor (DOL).

The safety and health function impacts virtually every function within a company or organization, and the safety and health professional does not work solely within a safety and health cocoon. There are safety and health components involved in virtually every aspect of every issue that arises within the operations. Each and every law or regulation has a direct or indirect impact on the safety and health function. Safety and health impact production, quality, environment, human resources, engineering, and virtually every department or function in any operation. The primary reason for this integration by the safety and health function is the interaction with the personnel within and outside of the operations.

The safety and health function has expanded from our early roots based in compliance to become an integral function within many companies and organizations. Many safety and health professionals, in addition to the primary safety and health function, also possess responsibilities for other such functions as workers' compensation, security, environmental, as well as many other varied functions. No matter how the safety and health position is structured within the company or organization, the one common thread among and between organizations and companies is that safety and health professionals impact people's lives. The safety and health function is often only visible when things go wrong. However, safety and health professionals often work daily to ensure a safe and healthful work environment and prepare programs, policies, and procedures that reduce the risk and avoid the potential accidents and the correlating costs.

Safety and health professionals should be cognizant that the safety and health function is no longer "OSHA centric," and the majority of many safety and health professionals' time involves activities and projects beyond achievement of basic compliance. Additionally, although compliance with the Occupational Safety and Health Administration (OSHA) standards is foundational and mandatory, safety and health professionals' programs and functions often far exceed the mandatory minimum requirements. The safety and health function has grown and expanded, being driven more by the cost savings and risk reduction for the company or organization than the fear of regulatory compliance penalties by OSHA. Most, if not all, of the companies and organizations employing a safety and health professional see the value of the safety and health professionals' experience and expertise in reducing a multitude of varying risks within the operations. Additionally, through the appropriate management of the safety and health function and correlating risk reduction, these companies and organizations incur cost savings and well as appropriate management of regulatory compliance.

Safety and health professionals should recognize the myriad other laws and regulations beyond the OSHA standards that directly or indirectly impact the safety and health function. On a federal level, safety and health professionals most often encounter the following:

Laws Enforced by EEOC

Title VII of the Civil Rights Act of 1964 (Title VII)

This law makes it illegal to discriminate against someone on the basis of race, color, religion, national origin, or sex. The law also makes it illegal to retaliate against a person because the person complained about discrimination, filed a charge of discrimination, or participated in an employment discrimination investigation or lawsuit. The law also requires that employers reasonably accommodate applicants' and employees' sincerely held religious practices, unless doing so would impose an undue hardship on the operation of the employer's business.

- ### *The Pregnancy Discrimination Act*
  This law amended Title VII to make it illegal to discriminate against a woman because of pregnancy, childbirth, or a medical condition related to pregnancy or childbirth. The law also makes it illegal to retaliate against a person because the person complained about discrimination, filed a charge of discrimination, or participated in an employment discrimination investigation or lawsuit.

The Equal Pay Act of 1963 (EPA)

This law makes it illegal to pay different wages to men and women if they perform equal work in the same workplace. The law also makes it illegal to retaliate against a person because the person complained about discrimination, filed a charge of discrimination, or participated in an employment discrimination investigation or lawsuit.

The Age Discrimination in Employment Act of 1967 (ADEA)

This law protects people who are 40 or older from discrimination because of age. The law also makes it illegal to retaliate against a person because the person complained about discrimination, filed a charge of discrimination, or participated in an employment discrimination investigation or lawsuit.

Title I of the Americans with Disabilities Act of 1990 (ADA)

This law makes it illegal to discriminate against a qualified person with a disability in the private sector and in state and local governments. The law also makes it illegal to retaliate against a person because the person complained about discrimination, filed a charge of discrimination, or participated in an employment discrimination investigation or lawsuit. The law also requires that employers reasonably accommodate the known physical or mental limitations of an otherwise qualified individual with a disability who is an applicant or employee, unless doing so would impose an undue hardship on the operation of the employer's business.

Sections 102 and 103 of the Civil Rights Act of 1991

Among other things, this law amends Title VII and the ADA to permit jury trials and compensatory and punitive damage awards in intentional discrimination cases.

---

* Equal Employment Opportunity Commission website: www.eeoc.gov.

Summary of the Major Laws of the Department of Labor
(Edited for the purposes of this text)

The Department of Labor (DOL) administers and enforces more than 180 federal laws. These mandates and the regulations that implement them cover many workplace activities for about 10 million employers and 125 million workers.

Following is a brief description of many of DOL's principal statutes most commonly applicable to businesses, job seekers, workers, retirees, contractors and grantees. This brief summary is intended to acquaint you with the major labor laws and not to offer a detailed exposition. For authoritative information and references to fuller descriptions on these laws, you should consult the statutes and regulations themselves.

Wages & Hours

*The **Fair Labor Standards Act (FLSA)** prescribes standards for wages and overtime pay, which affect most private and public employment. The act is administered by the Wage and Hour Division. It requires employers to pay covered employees who are not otherwise exempt at least the federal minimum wage and overtime pay of one-and-one-half-times the regular rate of pay. For nonagricultural operations, it restricts the hours that children under age 16 can work and forbids the employment of children under age 18 in certain jobs deemed too dangerous. For agricultural operations, it prohibits the employment of children under age 16 during school hours and in certain jobs deemed too dangerous.*

*The **Wage and Hour Division** also enforces the labor standards provisions of the **Immigration and Nationality Act (INA)** that apply to aliens authorized to work in the U.S. under certain nonimmigrant visa programs (H-1B, H-1B1, H-1C, H2A).*

Workplace Safety & Health

*The **Occupational Safety and Health (OSH)** Act is administered by the Occupational Safety and Health Administration (OSHA). Safety and health conditions in most private industries are regulated by OSHA or OSHA-approved state programs, which also cover public sector employers. Employers covered by the OSH Act must comply with the regulations and the safety and health standards promulgated by OSHA. Employers also have a general duty under the OSH Act to provide their employees with work and a workplace free from recognized, serious hazards. OSHA enforces the Act through workplace inspections and investigations. Compliance assistance and other cooperative programs are also available.*

Workers' Compensation

*The **Longshore and Harbor Workers' Compensation Act (LHWCA)**, administered by
The Office of Workers Compensation Programs (OWCP), provides for compensation and
medical care to certain maritime employees (including a longshore worker or other person
in longshore operations, and any harbor worker, including a ship repairer, shipbuilder, and
shipbreaker) and to qualified dependent survivors of such employees who are disabled or
die due to injuries that occur on the navigable waters of the United States, or in adjoining
areas customarily used in loading, unloading, repairing or building a vessel.*

*The **Energy Employees Occupational Illness Compensation Program Act (EEOICPA)** is
a compensation program that provides a lump-sum payment of $150,000 and prospective
medical benefits to employees (or certain of their survivors) of the Department of Energy and
its contractors and subcontractors as a result of cancer caused by exposure to radiation,
or certain illnesses caused by exposure to beryllium or silica incurred in the performance
of duty, as well as for payment of a lump-sum of $50,000 and prospective medical benefits
to individuals (or certain of their survivors) determined by the Department of Justice to be
eligible for compensation as uranium workers under section 5 of the **Radiation Exposure
Compensation Act (RECA)**.*

*The **Federal Employees' Compensation Act (FECA)**, 5 U.S.C. 8101 et seq., establishes a
comprehensive and exclusive workers' compensation program which pays compensation
for the disability or death of a federal employee resulting from personal injury sustained
while in the performance of duty. The FECA, administered by OWCP, provides benefits
for wage loss compensation for total or partial disability, schedule awards for permanent
loss or loss of use of specified members of the body, related medical costs, and vocational
rehabilitation.*

*The **Black Lung Benefits Act (BLBA)** provides monthly cash payments and medical benefits
to coal miners totally disabled from pneumoconiosis ("black lung disease") arising from their
employment in the nation's coal mines. The statute also provides monthly benefits to a
deceased miner's survivors if the miner's death was due to black lung disease.*

Employee Benefit Security

*The **Employee Retirement Income Security Act (ERISA)** regulates employers who offer
pension or welfare benefit plans for their employees. Title I of ERISA is administered
by the Employee Benefits Security Administration (EBSA) (formerly the Pension and
Welfare Benefits Administration) and imposes a wide range of fiduciary, disclosure and
reporting requirements on fiduciaries of pension and welfare benefit plans and on others
having dealings with these plans. These provisions preempt many similar state laws.
Under Title IV, certain employers and plan administrators must fund an insurance
system to protect certain kinds of retirement benefits, with premiums paid to the federal
government's Pension Benefit Guaranty Corporation (PBGC). EBSA also administers
reporting requirements for continuation of health-care provisions, required under the
Comprehensive Omnibus Budget Reconciliation Act of 1985*

*(COBRA) and the health care portability requirements on group plans under the **Health Insurance Portability and Accountability Act (HIPAA)**.*

Unions & Their Members

*The **Labor-Management Reporting and Disclosure Act (LMRDA)** of 1959 (also known as the Landrum-Griffin Act) deals with the relationship between a union and its members. It protects union funds and promotes union democracy by requiring labor organizations to file annual financial reports, by requiring union officials, employers, and labor consultants to file reports regarding certain labor relations practices, and by establishing standards for the election of union officers. The act is administered by the **Office of Labor-Management Standards (OLMS)**.*

Employee Protection

*Most labor and public safety laws and many environmental laws mandate whistleblower protections for employees who complain about violations of the law by their employers. Remedies can include job reinstatement and payment of back wages. **OSHA** enforces the whistleblower protections in most laws.*

Uniformed Services Employment and Reemployment Rights Act

*Certain persons who serve in the armed forces have a right to reemployment with the employer they were with when they entered service. This includes those called up from the reserves or National Guard. These rights are administered by the **Veterans' Employment and Training Service (VETS)**.*

Employee Polygraph Protection Act

*This law bars most employers from using lie detectors on employees, but permits polygraph tests only in limited circumstances. It is administered by the **Wage and Hour Division**.*

Garnishment of Wages

*Garnishment of employee wages by employers is regulated under the **Consumer Credit Protection Act (CPCA)** which is administered by the <u>Wage and Hour Division</u>.*

The Family and Medical Leave Act

*Administered by the <u>Wage and Hour Division</u>, the **Family and Medical Leave Act (FMLA)** requires employers of 50 or more employees to give up to 12 weeks of unpaid, job-protected leave to eligible employees for the birth or adoption of a child or for the serious illness of the employee or a spouse, child or parent.*

Veterans' Preference

*Veterans and other eligible persons have special employment rights with the federal government. **They are provided preference in initial hiring and protection in reductions in force.** Claims of violation of these rights are investigated by the <u>Veterans' Employment and Training Service (VETS)</u>.*

Government Contracts, Grants, or Financial Aid

Recipients of government contracts, grants or financial aid are subject to wage, hour, benefits, and safety and health standards under:

- *The **Davis-Bacon Act**, which requires payment of prevailing wages and benefits to employees of contractors engaged in federal government construction projects;*

- *The **McNamara-O'Hara Service Contract Act**, which sets wage rates and other labor standards for employees of contractors furnishing services to the federal government;*

- *The **Walsh-Healey Public Contracts Act**, which requires payment of minimum wages and other labor standards by contractors providing materials and supplies to the federal government.*

Administration and enforcement of these laws are by The <u>Wage and Hour Division</u>. The <u>Office of Federal Contract Compliance Programs (OFCCP)</u> administers and enforces three federal contract-based civil rights laws that require most federal contractors and subcontractors, as well as federally assisted construction contractors, to provide equal employment opportunity. The <u>Office of the Assistant Secretary for Administration and Management's (OASAM)</u> Civil Rights Center administers and enforces several federal assistance based civil rights laws *requiring recipients of federal financial assistance from Department of Labor to provide equal opportunity.*

Migrant & Seasonal Agricultural Workers

The Migrant and Seasonal Agricultural Worker Protection Act (MSPA) *regulates the hiring and employment activities of agricultural employers, farm labor contractors, and associations using migrant and seasonal agricultural workers. The Act prescribes wage protections, housing and transportation safety standards, farm labor contractor registration requirements, and disclosure requirements. The* <u>Wage and Hour Division</u> *administers this law.*

The Fair Labor Standards Act (FLSA) *exempts agricultural workers from overtime premium pay, but requires the payment of the minimum wage to workers employed on larger farms (farms employing more than approximately seven full-time workers. The Act has special child-labor regulations that apply to agricultural employment; children under 16 are forbidden to work during school hours and in certain jobs deemed too dangerous. Children employed on their families' farms are exempt from these regulations. The* <u>Wage and Hour Division</u> *administers this law. OSHA also has special safety and health standards that may apply to agricultural operations.*

The Immigration and Nationality Act (INA) *requires employers who want to use foreign temporary workers on H-2A visas to get a labor certificate from the* <u>Employment and Training Administration</u> *certifying that there are not sufficient, able, willing and qualified U.S. workers available to do the work. The labor standards protections of the H-2A program are enforced by The* <u>Wage and Hour Division</u>.

Mine Safety & Health

The Federal Mine Safety and Health Act of 1977 *(Mine Act) covers all people who work on mine property.* <u>**The Mine Safety and Health Administration (MSHA)**</u> *administers this Act.*

The Mine Act *holds mine operators responsible for the safety and health of miners; provides for the setting of mandatory safety and health standards, mandates miners' training requirements; prescribes penalties for violations; and enables inspectors to close dangerous mines. The safety and health standards address numerous hazards including roof falls, flammable and explosive gases, fire, electricity, equipment rollovers and maintenance, airborne contaminants, noise, and respirable dust. MSHA enforces safety and health requirements at more than 13,000 mines, investigates mine accidents, and offers mine operators training, technical and compliance assistance.*

Construction

*Several agencies administer programs related solely to the construction industry. OSHA has special occupational safety and health standards for construction; The Wage and Hour Division, under Davis-Bacon and related acts, requires payment of prevailing wages and benefits; The Office of Federal Contract Compliance Programs enforces **Executive Order 11246**, which requires federal construction contractors and subcontractors, as well as federally assisted construction contractors, to provide equal employment opportunity; the anti-kickback section of the **Copeland Act** precludes a federal contractor from inducing any employee to sacrifice any part of the compensation required.*

Transportation

*Most laws with labor provisions regulating the transportation industry are administered by agencies outside the Department of Labor. However, longshoring and maritime industry safety and health standards are issued and enforced by OSHA. The **Longshoring and Harbor Workers' Compensation Act (LHWCA)**, requires employers to assure that workers' compensation is funded and available to eligible employees. In addition, the rights of employees in the mass transit industry are protected when federal funds are used to acquire, improve, or operate a transit system. Under the **Federal Transit law**, the Department of Labor is responsible for approving employee protection arrangements before the department of Transportation can release funds to grantees.*

Plant Closings & *Layoffs*

*Such occurrences may be subject to the **Worker Adjustment and Retraining Notification Act (WARN)**. WARN offers employees early warning of impending layoffs or plant closings. The Employment and Training Administration (ETA) provides information to the public on WARN, though neither ETA nor the Department of Labor has administrative responsibility for the statute, which is enforced through private action in the federal courts.*

*This law makes it illegal to discriminate against a qualified person with a disability in the federal government. The law also makes it illegal to retaliate against a person because the person complained about discrimination, filed a charge of discrimination, or participated in an employment discrimination investigation or lawsuit. The law also requires that employers reasonably accommodate the known physical or mental limitations of an otherwise **qualified individual with a disability who is an applicant or employee**, unless doing so would impose an undue hardship on the operation of the employer's business.*

The Genetic Information Nondiscrimination Act of 2008 (GINA)

Effective - November 21, 2009.

This law makes it illegal to discriminate against employees or applicants because of genetic information. Genetic information includes information about an individual's genetic tests and the genetic tests of an individual's family members, as well as information about any disease, disorder or condition of an individual's family members (i.e. an individual's family medical history). The law also makes it illegal to retaliate against a person because the person complained about discrimination, filed a charge of discrimination, or participated in an employment discrimination investigation or lawsuit.

---

* U.S. Department of Labor website: www.dol.gov.

Although safety and health professionals may not encounter these federal laws or correlating state and local laws on a daily basis, it is essential that the safety and health professional be cognizant of these laws and be able to identify the possible interaction or impact of the law on the specific issue or situation. Given the interaction of safety and health within the organization and the various day-to-day issues, safety and health professionals can easily "step on their tail" in the decision-making process if evaluating the issue only from a safety and health perspective. Prudent safety and health professionals should analyze the potential impact of any and all correlating federal, state, and local laws and regulations before rendering any decision.

DISCUSSION QUESTIONS

1. How do the Americans with Disabilities Act (ADA) and Americans with Disabilities Act Amendments Act (ADAAA) interact with the safety and health function?
2. What is Genetic Information Nondescrimination Act (GINA) and how could this law potentially impact the safety and health function?
3. How does Title VII of the Civil Rights Act potentially impact the safety and health function?
4. Identify and discuss one law within the jurisdiction of the EEOC and identify the potential interaction with the safety and health function.

12 ADA and the Impact on Safety and Health

We are all born equal—equally helpless and equally indebted to others for whatever our survival turns out to be worth.

—**Cullen Hightower**

The hole and the patch should be commensurate.

—**Thomas Jefferson**

LEARNING OBJECTIVES

1. Identify and analyze the five titles within the ADA.
2. Assess and analyze the areas in which the Americans with Disabilities Act (ADA) and Americans with Disabilities Act Amendments Act (ADAAA) impact the safety and health function.
3. Analyze the test for qualification as being disabled under the ADA.
4. Identify and assess the potential legal and ethical issues created under the ADA and ADAAA for safety and health professionals.

One of the more extensive laws that can have a definite impact on the safety and health function is the Americans with Disabilities Act (ADA). In short, the ADA prohibits discriminating against qualified individuals with physical or mental disabilities in all employment settings. For safety and health professionals, workers' compensation, restricted duty programs, facility modifications, training, and other safety and health functions are often areas where the safety and health function and the ADA may intersect and create duties and responsibilities for the safety and health professional as well as potential liabilities for the company or organization.

Safety and health professionals should keep in mind that the ADA is an extensive law, and there are interpretations and decisions made in the court and agency virtually every day. A prudent safety and health professional does not need to be an expert in the ADA; however, he or she should be aware and cognizant of the requirements of the ADA and be able to recognize when the ADA is applicable to the safety situation.

It is important that safety and health professionals possess a base-level knowledge of the ADA, as well as key areas in which the ADA and safety function may intersect in order to be able to recognize when the ADA may be applicable to the situation. The agency responsible for enforcement is the Equal Employment

Opportunity Commission (EEOC) on the federal level, and information can be found on its website: www.eeoc.gov. Safety and health professionals should also be aware that individual states may also possess laws that parallel or more stringent than the federal ADA.

Additionally, it is vital that the safety and health professional acquires all of the facts before making a decision that may violate a qualified individual with a disability's rights under the ADA. Prudent safety and health professionals should gather all of the facts, document all aspects of the interaction, acquire assistance from human resources, EEO, or legal counsel, and review thoroughly prior to making any decisions involving the ADA. As noted in Chapter 9, safety and health professionals should refrain from making a snap decision, especially if the issue possesses possible ADA implications.

From a broad prospective, the ADA is divided into five titles, and all titles possess the potential of substantially impacting the safety and health function in covered public or private sector organizations. Title I contains the employment provisions that protect all individuals with disabilities who are in the United States, regardless of their national origin or immigration status. Title II prohibits discriminating against qualified individuals with disabilities or excluding them from the services, programs, or activities provided by public entities. Title II contains the transportation provisions of the act. Title III, entitled "Public Accommodations," requires that goods, services, privileges, advantages, and facilities of any public place be offered "in the most integrated setting appropriate to the needs of the individual."*

Title IV also covers transportation offered by private entities and addresses telecommunications. Title IV requires that telephone companies provide telecommunication relay services, and that public service television announcements that are produced or funded with federal money include closed caption. Title V includes the miscellaneous provisions. This title notes that the ADA does not limit or invalidate other federal and state laws providing equal or greater protection for the rights of individuals with disabilities, and addresses related insurance, alternate dispute, and congressional coverage issues.

Title I also prohibits covered employers from discriminating against a "qualified individual with a disability" with regard to job applications, hiring, advancement, discharge, compensation, training, and other terms, conditions, and privileges of employment.†

Section 101(8) defines a "qualified individual with a disability" as any person who, with or without reasonable accommodation, can perform the essential functions of the employment position that such individual holds or desires. Consideration shall be given to the employer's judgment as to what functions of a job are essential, and if an employer has prepared a written description before advertising or interviewing applicants for the job, this description shall be considered evidence of the essential function of the job.‡

---

* ADA Section 305.
† ADA Section 102(a); 42 USC Section 12112.
‡ ADA Section 101(8).

The Equal Employment Opportunity Commission (EEOC) provides additional clarification of this definition by stating, "An individual with a disability who satisfies the requisite skill, experience and educational requirements of the employment position such individual holds or desires, and who, with or without reasonable accommodation, can perform the essential functions of such position."*

Congress did not provide a specific list of disabilities covered under the ADA because "of the difficulty of ensuring the comprehensiveness of such a list."† Under the ADA, an individual has a disability if he or she:

- Possesses a physical or mental impairment that substantially limits one or more of the major life activities of such individual
- Possesses a record of such an impairment
- Is regarded as having such an impairment.‡

For an individual to be considered disabled under the ADA, the physical or mental impairment must limit one or more "major life activities." Under the U.S. Justice Department's regulation issued for Section 504 of the Rehabilitation Act, "major life activities" are defined as "functions such as caring for one's self, performing manual tasks, walking, seeing, hearing, speaking, breathing, learning and working."§ Congress clearly intended to have the term *disability* broadly construed. However, this definition does not include simple physical characteristics, nor limitations based on environmental, cultural, or economic disadvantages.¶

The second prong of this definition is "a record of such an impairment disability." The Senate Report and the House Judiciary Committee Report each stated:

> This provision is included in the definition in part to protect individuals who have recovered from a physical or mental impairment which previously limited them in a major life activity. Discrimination on the basis of such a past impairment would be prohibited under this legislation. Frequently occurring examples of the first group (i.e., those who have a history of an impairment) are people with histories of mental or emotional illness, heart disease or cancer; examples of the second group (i.e., those who have been misclassified as having an impairment) are people who have been misclassified as mentally retarded.**

The third prong of the statutory definition of a disability extends coverage to individuals who are "being regarded as having a disability." The ADA has adopted the same "regarded as" test that is used in Section 504 of the Rehabilitation Act:

> "Is regarded as having an impairment" means (A) has a physical or mental impairment that does not substantially limit major life activities but is treated ... as constituting

---

* EEOC Interpretive Rules, 56 Fed. Reg. 35 (July 26, 1991).
† 42 Fed. Reg. 22686 (May 4, 1977); S. Rep. 101-116; H. Rep. 101-485, Part 2, 51.
‡ Subtitle A, §3(2). The ADA departed from the Rehabilitation Act of 1973 and other legislation in using the term *disability* rather than *handicap*.
§ 28 CFR §41.31. This provision is adopted by and reiterated in the Senate Report on page 22.
¶ See *Jasany v. U.S. Postal Service*, 755 F.2d 1244 (6th Cir. 1985).
** S. Rep. 101-116, 23; H. Rep. 101-485, Part 2, 52–53.

such a limitation; (B) has a physical or mental impairment that substantially limits major life activities only as a result of the attitudes of others toward such impairment; (C) has none of the impairments defined (in the impairment paragraph of the Department of Justice regulations) but is treated ... as having such an impairment.[*]

Under the EEOC's regulations, this third prong covers three classes of individuals:

1. Persons who have physical or mental impairments that do not limit a major life activity but who are nevertheless perceived by covered entities (employers, places of public accommodation) as having such limitations. (For example, an employee with controlled high blood pressure that is not, in fact, substantially limited is reassigned to less strenuous work because of his employer's unsubstantiated fear that the individual will suffer a heart attack if he continues to perform strenuous work. Such a person would be regarded as disabled.)[†]

2. Persons who have physical or mental impairments that substantially limit a major life activity only because of a perception that the impairment causes such a limitation. (For example, an employee has a condition that periodically causes an involuntary jerk of the head, but no limitations on his major life activities. If his employer discriminates against him because of the negative reaction of customers, the employer would be regarding him as disabled and acting on the basis of that perceived disability.)[‡]

3. Persons who do not have a physical or mental impairment, but are treated as having a substantially limiting impairment. (For example, a company discharges an employee based on a rumor that the employee is HIV-positive. Even though the rumor is totally false and the employee has no impairment, the company would nevertheless be in violation of the ADA.)[§]

Thus, a "qualified individual with a disability" under the ADA is any individual who can perform the essential or vital functions of a particular job with or without the employer accommodating the particular disability. The employer is provided the opportunity to determine the "essential functions" of the particular job before offering the position through the development of a written job description. This written job description will be considered evidence to which functions of the particular job are essential and which are peripheral. In deciding the essential functions of a particular position, the EEOC will consider the employer's judgment, whether the written job description was developed prior to advertising or beginning the interview process, the amount of time spent performing the job, the past and current

---

[*] 45 CFR 84.3(j)(2)(iv), quoted from H. Rep. 101-485, Part 3, 29; S. Rep. 101-116, 23; H. Rep. 101-485, Part 2, 53; also see *School Board of Nassau County, Florida v. Arline*, 107 S. Ct. 1123 (1987) (leading case).

[†] EEOC Interpretive Guidelines, 56 Fed. Reg. 35,742 (July 26, 1991).

[‡] S. Comm. on Lab. and Hum. Resources Rep. at 24; H. Comm. on Educ. and Lab. Rep. at 53; H. Comm. on Jud. Rep. at 30–31.

[§] 29 CFR §1630.2(1).

experience of the individual to be hired, relevant collective bargaining agreements, and other factors.*

The EEOC defines the term *essential function* of a job as meaning "primary job duties that are intrinsic to the employment position the individual holds or desires" and precludes any marginal or peripheral functions that may be incidental to the primary job function.† The factors provided by the EEOC in evaluating the essential functions of a particular job include the reason that the position exists, the number of employees available, and the degree of specialization required to perform the job.‡ This determination is especially important to safety and health professionals who may be required to develop the written job descriptions or to determine the essential functions of a given position.

Safety and health professionals should recognize the important and pertinent issue of "direct threat" to the safety and health of the individual or others. Safety and health professionals should recognize this situation when the treatment of the disabled individual, as a matter of fact or due to prejudice, is believed to be a direct threat to the safety and health of themselves or others in the workplace. This sensitive issue often places the burden directly on the shoulders of the safety and health professional to evaluate and render a decision that will impact not only the individual with a disability but also the company or organization. To address this issue, the ADA provides that any individual who poses a direct threat to the health and safety of others that cannot be eliminated by reasonable accommodation may be disqualified from the particular job.§ The term *direct threat* to others is defined by the EEOC as creating "a significant risk of substantial harm to the health and safety of the individual or others that cannot be eliminated by reasonable accommodation."¶ The determining factors that safety and health professionals should consider in making this determination include the duration of the risk, the nature and severity of the potential harm, and the likelihood that the potential harm will occur.**

Additionally, safety and health professionals should consider the EEOC's Interpretive Guidelines, which state:

> [If] an individual poses a direct threat as a result of a disability, the employer must determine whether a reasonable accommodation would either eliminate the risk or reduce it to an acceptable level. If no accommodation exists that would either eliminate the risk or reduce the risk, the employer may refuse to hire an applicant or may discharge an employee who poses a direct threat.††

Safety and health professionals should also note that Title I additionally provides that if an employer does not make reasonable accommodations for the known limitations of a qualified individual with disabilities, it is considered to be discrimination.

---

* ADA, Title I, Section 101(8).
† EEOC Interpretive Rules, supra note 11.
‡ Id.
§ ADA Section 103(b).
¶ EEOC Interpretive Guidelines, supra, note 11.
** Id.
†† 56 Fed. Reg. 35,745 (July 26, 1991); also see *Davis v. Meese*, 692 F. Supp. 505 (E.D. Pa. 1988) (Rehabilitation Act decision).

Only if the employer can prove that providing the accommodation would place an undue hardship on the operation of the employer's business can discrimination be disproved. Section 101(9) defines a "reasonable accommodation" as:

(a) making existing facilities used by employees readily accessible to and usable by the qualified individual with a disability and includes:

(b) job restriction, part-time or modified work schedules, reassignment to a vacant position, acquisition or modification of equipment or devices, appropriate adjustments or modification of examinations, training materials, or policies, the provisions of qualified readers or interpreters and other similar accommodations for ... the QID (qualified individual with a disability).[*]

The EEOC further defines "reasonable accommodation" as:

1. Any modification or adjustment to a job application process that enables a qualified individual with a disability to be considered for the position such qualified individual with a disability desires, and which will not impose an undue hardship on the ... business; or

2. Any modification or adjustment to the work environment, or to the manner or circumstances which the position held or desired is customarily performed, that enables the qualified individual with a disability to perform the essential functions of that position and which will not impose an undue hardship on the ... business; or

3. Any modification or adjustment that enables the qualified individual with a disability to enjoy the same benefits and privileges of employment that other employees enjoy and does not impose an undue hardship on the ... business.[†]

Safety and health professionals should be aware that the company or organization would be required to make reasonable accommodations for any and all known physical or mental limitations of the qualified individual with a disability, unless the employer can demonstrate that the accommodations would impose an "undue hardship" on the business, or that the particular disability directly affects the safety and health of that individual or others. Safety professionals should also be aware that included under this section is the prohibition against the use of qualification standards, employment tests, and other selection criteria that can be used to screen out individuals with disabilities, unless the employer can demonstrate that the procedure is directly related to the job function. In addition to the modifications to facilities, work schedules, equipment, and training programs, the company or organization is required to initiate an "informal interactive (communication) process" with the qualified individual to promote voluntary disclosure of his or her specific limitations and restrictions to enable the employer to make appropriate accommodations that will compensate for the limitation.[‡]

Safety and health professionals should be aware that Section 101(10)(a) defines "undue hardship" as "an action requiring significant difficulty or expense," when considered in light of the following factors:

---

[*] ADA Section 101(9).

[†] EEOC Interpretive Guidelines, supra note 11.

[‡] Id.

- The nature and cost of the accommodation
- The overall financial resources and workforce of the facility involved
- The overall financial resources, number of employees, and structure of the parent entity
- The type of operation, including the composition and function of the work-force, the administration, and the fiscal relationship between the entity and the parent[*]

Of particular importance to safety professionals is Section 102(c)(1) of the ADA. This section prohibits discrimination through medical screening, employment inquiries, and similar scrutiny. Safety professionals should be aware that underlying this section was Congress's conclusion that information obtained from employment applications and interviews "was often used to exclude individuals with disabilities—particularly those with so-called hidden disabilities such as epilepsy, diabetes, emotional illness, heart disease and cancer—before their ability to perform the job was even evaluated."[†]

Additionally, under Section 102(c)(2), safety and health professionals should be aware that conducting preemployment physical examinations of applicants and asking prospective employees if they are qualified individuals with disabilities is prohibited. Employers are further prohibited from inquiring as to the nature or severity of the disability, even if the disability is visible or obvious. Safety and health professionals should be aware that individuals may ask whether any candidates for transfer or promotion who have a known disability can perform the required tasks of the new position if the tasks are job related and consistent with business necessity. An employer is also permitted to inquire about the applicant's ability to perform the essential job functions prior to employment. The employer should use the written job description as evidence of the essential functions of the position.[‡]

Safety and health professionals may require medical examinations of employees only if the medical examination is specifically job related and is consistent with business necessity. Medical examinations are permitted only after the applicant with a disability has been offered the job position. The medical examination may be given before the applicant starts the particular job, and the job offer may be contingent upon the results of the medical examination if all employees are subject to the medical examinations and information obtained from the medical examination is maintained in separate, confidential medical files. Employers are permitted to conduct voluntary medical examinations for current employees as part of an ongoing medical health program, but again, the medical files must be maintained separately and in a confidential manner.[§] The ADA does not prohibit safety and health professionals or their medical staff from making inquiries or requiring medical or "fit for duty" examinations when there is a need to determine whether or not an employee is still able to

---

[*] See *Gruegging v. Burke*, 48 Fair Empl. Prac. Cas. (BNA) 140 (D.D.C. 1987); *Bento v. ITO Corp.*, 599 F. Supp. 731 (D.R.I. 1984).
[†] S. Comm. on Lab. and Hum. Resources Rep. at 38; H. Comm. on Jud. Rep. at 42.
[‡] ADA. Title I, Section 102(c)(2).
[§] ADA Section 102(c)(2)(A).

perform the essential functions of the job, or where periodic physical examinations are required by medical standards or federal, state, or local law.*

Another area of particular importance for safety and health professionals is the area of controlled substance testing. Under the ADA, the employer is permitted to test job applicants for alcohol and controlled substances prior to an offer of employment under Section 104(d). The testing procedure for alcohol and illegal drug use is not considered a medical examination as defined under the ADA. Employers may additionally prohibit the use of alcohol and illegal drugs in the workplace and may require that employees not be under the influence while on the job. Employers are permitted to test current employees for alcohol and controlled substance use in the workplace to the limits permitted by current federal and state law. The ADA requires all employers to conform to the requirements of the Drug-Free Workplace Act of 1988. Thus, safety and health professionals should be aware that most existing pre-employment and postemployment alcohol and controlled substance programs that are not part of the preemployment medical examination or ongoing medical screening program will be permitted in their current form.† Individual employees who choose to use alcohol and illegal drugs are afforded no protection under the ADA. However, employees who have successfully completed a supervised rehabilitation program and are no longer using or addicted are offered the protection of a qualified individual with a disability under the ADA.‡

Title II of the ADA is designed to prohibit discrimination against disabled individuals by public entities. This title covers the provision of services, programs, activities, and employment by public entities. A public entity under Title II includes:

- A state or local government
- Any department, agency, special purpose district, or other instrumentality of a state or local government
- The National Railroad Passenger Corporation (Amtrak), and any commuter authority as this term is defined in Section 103(8) of the Rail Passenger Service Act§

Although limited in the applicability for public sector safety and health professionals, Title II of the ADA prohibits discrimination in the area of ground transportation, including buses, taxis, trains, and limousines. Air transportation is excluded from the ADA but is covered under the Air Carriers Access Act.[41] Covered organizations may be affected in the purchasing or leasing of new vehicles and in other areas, such as the transfer of disabled individuals to the hospital or other facilities. Title II requires covered public entities to make sure that new vehicles are accessible to and usable by the qualified individual, including individuals in wheelchairs. Thus, vehicles must be equipped with lifts, ramps, wheelchair space, and other modifications unless the

---

* EEOC Interpretive Guidelines, 56 Fed. Reg. 35,751 (July 26, 1991). Federally mandated periodic examinations include such laws as the Rehabilitation Act, Occupational Safety and Health Act, Federal Coal Mine Health Act, and numerous transportation laws.
† ADA Section 102(c).
‡ ADA Section 511(b).
§ ADA Section 201(1).

covered public entity can justify that such equipment is unavailable despite a good faith effort to purchase or acquire this equipment. Covered organizations may want to consider alternative methods to accommodate the qualified individual, such as use of ambulance services or other alternatives.

Title III of the ADA builds upon the foundation established by the Architectural Barriers Act and the Rehabilitation Act. This title basically extends the prohibitions that currently exist against the prohibition discrimination to apply to all privately operated public accommodations. Title III focuses on the accommodations in public facilities, including such covered entities as retail stores, law offices, medical facilities, and other public areas. This section requires that goods, services, and facilities of any public place provide "the most integrated setting appropriate to the needs of the (qualified individual with a disability)" except where that individual may pose a direct threat to the safety and health of others that cannot be eliminated through modification of company procedures, practices, or policies. Prohibited discrimination under this section includes prejudice or bias against the individual with a disability in the "full and equal enjoyment" of these services and facilities.[*]

The ADA makes it unlawful for public accommodations not to remove architectural and communication barriers from existing facilities or transportation barriers from vehicles "where such removal is readily achievable."[†] This statutory language is defined as "easily accomplished and able to be carried out without much difficulty or expense,"[‡] for example, moving shelves to widen an aisle, lowering shelves to permit access, etc. The ADA also requires that when a commercial facility or other public accommodation is undergoing a modification that affects the access to a primary function area, specific alterations must be made to afford accessibility to the qualified individual with a disability.

Title III also requires that "auxiliary aids and services" be provided for the qualified individual with a disability, including, but not limited to, interpreters, readers, amplifiers, and other devices (not limited or specified under the ADA) to provide that individual with an equal opportunity for employment, promotion, etc.[§] Congress did, however, provide that auxiliary aids and services do not need to be offered to customers, clients, and other members of the public if the auxiliary aid or service creates an undue hardship on the business. Safety and loss prevention professionals may want to consider alternative methods of accommodating the qualified individual with a disability. This section also addresses the modification of existing facilities to provide access to the individual, and requires that all new facilities be readily accessible and usable by the individual.

Safety and health professionals should be aware of Title IV; however, there is limited applicability for most private sector safety professionals. Title IV requires all telephone companies to provide "telecommunications relay service" to aid the hearing- and speech-impaired individuals. The Federal Communication Commission issued a regulation requiring the implementation of this requirement

---

[*] ADA Section 302.
[†] ADA Section 302(b)(2)(A)(iv).
[‡] ADA Section 301(9).
[§] ADA Section 3(1).

by July 26, 1992, and also established guidelines for compliance. This section also requires that all public service programs and announcements funded with federal monies be equipped with closed caption for the hearing impaired.*

Safety and health professionals should be aware that Title V assures that the ADA does not limit or invalidate other federal or state laws that provide equal or greater protection for the rights of individuals with disabilities. Thus, safety and health professionals should also be aware of any individual state laws or regulations addressing the same or similar areas as the ADA.

When enacting the ADA, safety and health professionals should be aware that Congress expressed its concern that sexual preferences could be perceived as a protected characteristic under the ADA or that the courts could expand ADA's coverage beyond Congress's intent. Accordingly, Congress included Section 511(b), which contains an expansive list of conditions that are not to be considered within the ADA's definition of disability. This list includes individuals such as transvestites, homosexuals, and bisexuals. Additionally, the conditions of transsexualism, pedophilia, exhibitionism, voyeurism, gender identity disorders not resulting from physical impairment, and other sexual behavior disorders are not considered as a qualified disability under the ADA. Compulsive gambling, kleptomania, pyromania, and psychoactive substance use disorders (from current illegal drug use) are also not afforded protection under the ADA.†

Safety and health professionals should be aware that all individuals associated with or having a relationship to the qualified individual with a disability are extended protection under this section of the ADA. This inclusion is unlimited in nature, including family members, individuals living together, and an unspecified number of others.‡ The ADA extends coverage to all individuals, legal or illegal, documented or undocumented, living within the boundaries of the United States, regardless of their status.§ Under Section 102(b)(4), unlawful discrimination includes "excluding or otherwise denying equal jobs or benefits to a qualified individual because of the known disability of the individual with whom the qualified individual is known to have a relationship or association."¶ Therefore, the protections afforded under this section are not limited to only familial relationships. There appears to be no limits regarding the kinds of relationships or associations that are afforded protection. Of particular note is the inclusion of unmarried partners of persons with AIDS or other qualified disabilities.**

As with the OSH Act, the ADA requires that employers post notices of the pertinent provisions of the ADA in an accessible format in a conspicuous location within

---

* Report of the House Committee on Energy and Commerce on the Americans with Disabilities Act of 1990, H.R. Rep. No. 485, 101st Cong., 2d Sess. (1990) (hereinafter cited as H. Comm. on Energy and Comm. Rep.); H. Comm. on Educ. and Lab. Rep., supra; S. Comm. on Lab. and Hum. Resources Rep.
† ADA §§511(a), (b); 508. There is some indication that many of the conditions excluded from the disability classification under the ADA may be considered a covered handicap under the Rehabilitation Act. See *Rezza v. Dept. of Justice*, 46 Fair Empl. Prac. Cas. (BNA) 1336 (E.D. Pa. 1988) (compulsive gambling); *Fields v. Lyng*, 48 Fair Empl. Prac. Cas. (BNA) 1037 (D. Md. 1988) (kleptomania).
‡ ADA Sections 102 and 302.
§ H. Rep. 101-485, Part 2, 51.
¶ ADA Section 102.
** H. Rep. 101-485, Part 2, 61-62, 38-39.

the employer's facilities. A prudent safety and loss prevention professional may wish to provide additional notification on job applications and other pertinent documents.[*]

Under the ADA, safety and health professionals should be aware that it is unlawful for an employer to "discriminate on the basis of disability against a qualified individual with a disability" in all areas, including:

- Recruitment, advertising, and job application procedures
- Hiring, upgrading, promoting, awarding tenure, demotion, transfer, layoff, termination, the right to return from layoff, and rehiring
- Rate of pay or other forms of compensation and changes in compensation
- Job assignments, job classifications, organization structures, position descriptions, lines of progression, and seniority lists
- Leaves of absence, sick leave, or other leaves
- Fringe benefits available by virtue of employment, whether or not administered by the employer
- Selection and financial support for training, including apprenticeships, professional meetings, conferences, and other related activities, and selection for leave of absence to pursue training
- Activities sponsored by the employer, including social and recreational programs
- Any other term, condition, or privilege of employment[†]

The EEOC has also noted that it is "unlawful ... to participate in a contractual or other arrangement or relationship that has the effect of subjecting the covered entity's own qualified applicant or employee with a disability to discrimination." This prohibition includes referral agencies, labor unions (including collective bargaining agreements), insurance companies and others providing fringe benefits, and organizations providing training and apprenticeships.[‡]

Safety and health professionals should be aware that the ADA has the same enforcement and remedy scheme as Title VII of the Civil Rights Act of 1964, as amended by the Civil Rights Act of 1991. Compensatory and punitive damages (with upper limits) have been added as remedies in cases of intentional discrimination, and there is also a correlative right to a jury trial. Unlike Title VII, there is an exception when there is a good faith effort at reasonable accommodation.[§]

Safety and health professionals should be aware that the governing federal agency for the ADA is the Equal Employment Opportunity Commission. Enforcement of the ADA is also permitted by the attorney general or by private lawsuit. Remedies under these titles include the ordered modification of a facility, and civil penalties of up to $50,000 for the first violation and $100,000 for any subsequent violations. Section 505 permits reasonable attorney fees and litigation costs for the prevailing

---

[*] ADA Section 105.
[†] EEOC Interpretive Guidelines, EEOC, 1994.
[‡] Id.
[§] Civil Rights Act of 1992, Section 102.

party in an ADA action, but under Section 513, Congress encourages the use of arbitration to resolve disputes arising under the ADA.*

With the passage of the Civil Rights Act of 1991, the remedies provided under the ADA were modified. Employment discrimination (whether intentional or by practice) that has a discriminatory effect on qualified individuals may include hiring, reinstatement, promotion, back pay, front pay, reasonable accommodation, or other actions that will make an individual "whole." Payment of attorney fees, expert witness fees, and court fees are still permitted, and jury trials also allowed.

Compensatory and punitive damages are also made available if intentional discrimination is found. Damages may be available to compensate for actual monetary losses, future monetary losses, mental anguish, and inconvenience. Punitive damages are also available if an employer acted with malice or reckless indifference. The total amount of punitive and compensatory damages for future monetary loss and emotional injury for each individual is limited, and is based upon the size of the employer.

Although safety and health professionals are not expected to be ADA experts, it is important that they possess a grasp of the general requirements of the ADA as well as issues that may impact the safety and health function. As identified in Chapter 9, it is important for safety and health professionals to listen to their employees in order to identify potential ADA issues and acquire the appropriate guidance from human resources, legal counsel, or their company's ADA professionals.

DISCUSSION QUESTIONS

1. Identify the one title under the ADA that most impacts the safety and health function and explain why.
2. Explain how a safety and health professional may be involved in the assessment in addressing a request for reasonable accommodation.
3. Which federal agency possesses jurisdiction over the ADA? Identify where information can be located.
4. What is the test for an individual with a disability to qualify for protection under the ADA?

---

* ADA Sections 505 and 513.

13 Impact of Happiness on Safety

Happiness is good health and a bad memory.

—**Ingrid Bergman**

Happiness is not a goal, it is a by-product.

—**Eleanor Roosevelt**

LEARNING OBJECTIVES

1. Analyze and assess happiness as applicable to the safety and health function.
2. Analyze and assess the impact of happiness on safety and health performance.
3. Analyze and assess the things that can create happiness in the workplace.
4. Identify and analyze the benefits of creating a happy workplace.

Safety and health professionals for decades have been utilizing incentive programs in an attempt to positively motivate their employees to work safely. Safety and health incentive programs often span the spectrum from drawings to win a car to steak dinners for achieving a safety and health goal to more formalized "green stamp" programs to reward employees for good safety and health performance and permit employees to select individualized rewards. Although safety and health incentive programs are often utilized to motivate the workforce to achieve specific safety and health goals, safety and health professionals have found that this type of motivation can be effective; however, many safety and health professionals have found that incentive programs are not a substitute for a foundational safety and health program and the benefits can be short-lived. However, positive reinforcement of safety and health can be beneficial when properly designed and implemented to achieve short-term goals and objectives.

Historically, enforcement of safety and health rules and regulations on the shop floor has been through negative reinforcement, i.e., disciplinary action. Under the Occupational Safety and Health Act of 1970, employers and employees who were not in compliance received negative reinforcement in the form of citations and monetary penalties.[*] Negative reinforcement, in varying forms through individual company or

---

[*] Pub. L. 91-596, 84 Stat. 1590, 91st Congress, S.2193, December 29, 1970, as amended through January 1, 2004. Note: OSHA has only cited employers for noncompliance. To date, no individual employee has received a citation or monetary penalty.

organization's disciplinary policies and procedures, has been a basic staple through which safety and health compliance is achieved and maintained in most operations.

Throughout the decades, safety and health professionals have utilized a number of different programs and techniques through which to empower employees to take responsibility for their own and their fellow employees' safety and health in the workplace. Achieving this cultural shift in responsibility within the safety and health function requires change in process and thinking, an acceptance of responsibility, and the willingness to prioritize safety and health at the highest level. Although some companies and organizations have been able to achieve this cultural shift, most companies and organizations revert to the fallback position of negative reinforcement.

The safety and health function can be impacted not only by other laws and regulations, but also by other variables and influences far beyond the control of the safety and health professional. The safety and health function can be impacted by the profitability, or lack thereof, of the company, weather impacts on the operations, overtime requirements, and work hour reductions, to name a few, in addition to such broader activities as downsizing, layoffs, forced retirements, union organizing campaigns, and others. The common factors consistent among all of these variables and influences on the safety and health function are change and the individual and collective employee is not happy.

Is the happiness of your employees driven by genetics? Environment? Money? Other factors? Do employees work safer if they are happy? Are happy employees good for business? Are happy workers healthier? As identified by Ted Marusarz, the leader of Global Engagement and Culture, in a recent study on business financial health, "Understanding what drives employee behavior—in good times and in bad—is critical to business success.... All organizations face similar pressures. Companies that are successful at improving engagement in spite of these pressures are the ones that create an environment focused on key human capital elements. They may make adjustments to their engagement strategies, but they don't lose sight of their overall goals."*

First, safety and health professionals should consider what *happy* means within the context of their safety and health program and organization. "Happiness is a broad and subjective word, but a person's well-being includes the presence of positive emotions, like joy and interest, and the absence of negative emotions, like apathy and sadness."† "Happiness is not only a responsibility to ourselves, but also to our co-workers, who often rely on us to be steadfast and supportive. In addition, employee well-being affects the organization overall. Studies have shown that after controlling for age, gender, ethnicity, job tenure and educational attainment levels, psychological well-being still is significantly related to job performance."‡

For safety and health professionals exploring how to create a happy workforce and motivate their employees to work safely, one of the first questions to be asked by companies or organizations is, can money create happiness within the workforce, and if so, how much? According to the vast majority of the studies, money does not

---

* "Why Happy Employees Are Good for Business," Courtney Rubin, Inc.com/news/articles/2010/09/happy-employees are-good-for-business.
† "Happy Employees Are Critical for an Organization's Success," www.sciencedaily.com (February 4, 2009) (a study conducted at Kansas State University by Thomas Wright and Jon Wefald).
‡ Id.

equate to happiness for most workers. As identified earlier in this text, Maslow's hierarchy of physiological needs identifies safety and security as the top priority, with financial security, savings accounts, insurance policies, and related financial "safety net" protections being important components of this top priority, and this priority is often met by the company or organization.* If the employee is paid adequately and possesses the basic financial protections and insurances, more money often does not motivate the employee and create happiness within the workforce.

If money doesn't create happy employees, safety and health professionals should explore alternative methodologies through which to fulfill the needs of their employees. "When employees feel that the company takes their interest to heart, then the employee will take the company interest to heart."† "What business owners need to do is keep their promises and show compassion for their employees. So if you promise a pay raise, give it to them. If you promise resources to help them be more efficient at their job, give it to them. Find what's important to your employees and give it to them."‡

One of the first areas for consideration by safety and health professionals is a self-examination of each and every member of your management team. Bad bosses equate to an unhappy worker and potentially unsafe situations. Arguable, unhappy workers correlate to unsafe workers, which parallels increased workers' compensation, health insurance, and related costs. In a recent study, "only 36 percent of American workers reported they were happy at their jobs. That means that as many as 64 percent are unhappy at work—and their bosses appear largely to blame. In fact, 65 percent said a better boss would make them happier at work; only 35 percent chose a pay raise."§ Furthermore, the study found "in terms of the impact a boss has on employee health, 73 percent of those in their 20's and 30's said their health was at stake while only 40 percent of those 50 and older felt this way," and "when stress levels rise at work, 47 percent reported their boss does not stay calm and in control."¶

For safety and health professionals, what things can achieve happiness in your workforce and correlate to a safer and more healthful workplace? Although the things can vary depending upon the makeup, structure, location, and other variables of your workforce, as well as the situations, such as layoffs, cut hours, and other variables, below are basic concepts to consider in order to create and maintain happy workers in your workplace:

1. Management recognizes that employees are people first, workers second.**
2. "A workplace is far likelier to be a happy place when policies are in place and ensure that people regularly get acknowledgement and praise for a job well done."††

---

* See Maslow's hierarchy of needs, www.web.archives.org.
† "Make More Money by Making Your Employees Happy," Dr. Noelle Nelson, www.forbes.com (July 7, 2012).
‡ Id.
§ "Unhappy Workers Would Choose New Boss over More Pay," Laura Walter, www.ehstoday.com (October 17, 2012).
¶ Id.
** "What Top Companies Know: The 5 Basic Rules of Happy Employees," Derek Irvine, www.tlnt.com (January 11, 2013).
†† Id.

3. Empowered employees are happier employees. "Having meaningful impact on the world around you is actually a better predictor of happiness than other things."*
4. Remove the status quo. Job movement, job challenges, improvement, or the perception of improvement creates challenges and can eliminate job burnout.†
5. Offer challenging and good training programs.‡
6. Promote autonomy to employees and allow employees to focus on one job at a time.§
7. Offer small bonuses, incentives, and praise to employees.¶
8. Offer unique benefit packages meeting individual employees' needs.**
9. Promote work-life integration and balance.††
10. Provide a break from the routine for other activities.‡‡
11. Create and maintain a favorable working environment.
12. Consider "out of the box" ideas by your employees.
13. Give employees a say in decisions that impact their safety, health, job, or related issues.
14. Evaluate and eliminate negative working conditions. Every job is important.
15. Provide the tools to supervisors and manager to create a caring and supportive management team.

Creating and maintaining a positive environment with effective lines of communication and employee empowerment is setting the stage for a happy and safe working environment. As identified by E. Scott Geller, "companies need to learn from their employees because safety is best accomplished from the bottom up.... You will never be able to eliminate injuries, but you will get a lot closer to the source when you involve your employees."§§ Additionally, in a study conducted by Marshall University it was identified that "a more organic atmosphere is needed in order to provide more open verbal/non-verbal communications. The supervisor must take immediate action to remedy safety problems and positively reinforce examples of safe behavior. It is not what managers say that will matter to workers, it is what they do that communicates how important worker safety is to the organization."¶¶ The cost of creating and maintaining a safe and happy work environment does not appear to correlate to dollars but to the methodologies employed in effectively and efficiently communicating with employees and the methods through which employees are managed within the working environment. Empowerment of employees, positive

---

* Id.
† Id.
‡ "11 Things That Make Workers Happy," www.businessnews dily.com/3132/html.
§ Id.
¶ Id.
** Id.
†† Id., notes 10 and 14.
‡‡ Id., note 10.
§§ "The Psychology of Safety: How to Improve Behaviors and Attitudes," E.S. Geller, Chilton Book Co., Radnor, PA. (2000)
¶¶ "Job Satisfaction as Related to Safety Performance: A Case for a Manufacturing Firm," C.W. Kim, M.L. McInerney, and R.P. Alexander, *Coastal Business Journal*, 1(1). (2002)

reinforcement, providing challenges to employees, and effective training programs appear to be the first step in creating a happy atmosphere. How much does it cost to tell an employee he or she is doing a good job?

A recent study titled "Can Worker Safety Impact Customer Satisfaction?" "which was conducted by the National Safety Council (NSC) and published in the *Journal of Safety Research*, studied 821 employees at a Midwestern electric utility company's power delivery and customer care groups to consider how a company's safety climate and workplace injury statistics might impact customer satisfaction."

> The study examined work groups at the utility company that were responsible for customer-related functions including installation and service of distribution lines, meter reading, billing, safety, emergency services and more. Work units that had more employee injuries, researchers revealed, also had customers who were less satisfied with the service they received.
>
> Researchers considered that safe working environments may create other benefits related to quality of the work, which may subsequently impact customer satisfaction.
>
> "In an organization with a positive safety climate, where safety does not take a back seat to productivity, employees are likely to believe they have permission to do things right. Doing things right is a permeating value in a work unit that is likely to reach into several domains of work behavior, some of which influence the quality of work," the paper stated.
>
> The results show that customer satisfaction and a company's safety climate and injury rates were "significantly correlated." The researchers also suggested that the study results help make the business case for safety:
>
> "The results of this research bolster the business case for safety. They demonstrate that workplace safety is not simply an issue of doing the right thing or avoiding costs associated with lost-time injuries and related expenses. There are positive business outcomes to be gained in the form of improved customer satisfaction. The explanatory logic, although not proven conclusively here, is that a better safety environment produces spillover effects into the service environment," the study stated.[*]

DISCUSSION QUESTIONS

1. How can a happy workplace impact the safety and health function?
2. Does happiness impact injury and illness rates? Why or why not?
3. Can safety and health incentive programs create happiness in the workplace? Why or why not?
4. Can a "bad boss" impact the safety and health function? Why or why not?

---

[*] "Can Worker Safety Impact Customer Satisfaction?" www.ehstoday.com/safety (January 20, 2013).

14 Here Are Your OSHA Citations!

A moment of thinking is an hour in words.

—**Thomas Hood**

There isn't a plant or a business on earth that couldn't stand a few improvements—and be better for them. Someone is going to think of them. Why not beat the other fellow to it?

—**Roger W. Babson**

LEARNING OBJECTIVES

1. Acquire an understanding of the OSHA citation and penalty processes.
2. Acquire an understanding of the OSHA informal conference process.
3. Acquire an understanding of the OSHA/OSHRC appeals process.

One of the first "OMG" moments in the career of many safety and health professionals is when the Occupational Safety and Health Administration (OSHA) citations are received and opened. Safety and health professionals are often involved throughout the inspection process, including in the opening conference, the compliance inspection, and the closing conference; however, the actual opening of the envelop containing the codified proposed violations and the proposed monetary penalties can be a shock. The safety and health professional is usually aware of the potential violations and has often taken steps to correct the alleged violations identified during the compliance inspection; however, receipt of the citations has started the 15-working-day clock to appeal the citations. What can and should the safety and health professional recommend to the management team to address these alleged violations and proposed penalties?

One often overlooked option for safety and health professionals whose company or organization has received a citation is the informal conference. In essence, an informal conference is an opportunity to meet and discuss in a rather informal setting the proposed citations, proposed penalties, proposed abatement, and any or all other issues related to the inspection or citation with the area director for OSHA (or your state plan program). An informal conference also provides an opportunity to clarify any issues and provides an opportunity to discuss and negotiate possible settlement of the proposed citations prior to the filing of a formal appeal of the citations. However, safety and health professionals must be aware that any request for an informal conference must be initiated by the safety and health professional and must be conducted within the 15-working-day time period, but before filing a formal Notice of Contest.

OSHA encourages the use of informal conferences, and additionally, an informal conference can be very beneficial to safety professionals. The benefits to OSHA include the elimination of the need for the formal appeals process before the Occupational Safety and Health Review Commission (OSHRC), the subsequent litigation that can result, and the costs and manpower requirements involved in this type of litigation. The benefits to the safety and health professional and his or her organization can include resolution of the matter, minimized impact on the safety efforts, reduced litigation costs, and reduced manpower requirements. In essence, an informal conference, if the matters can be resolved, is a win-win for all involved parties.

Safety and health professionals are cautioned to carefully analyze the possible benefits as well as possible detriments when determining whether or not to request an informal conference. Generally, the safety and health professional, as well as the management team and legal counsel, should assess the number of citations, degree of the alleged violations, possible defenses to the alleged violation, proposed monetary penalties, potential criminal penalties, and other factors before requesting and participating in an informal conference. Generally, informal conferences are utilized as a settlement vehicle for citations containing a relatively small number of alleged violations, lower degrees of violations, and relatively smaller proposed monetary penalties. For citations containing willful violations, above relatively large proposed monetary penalties, or possessing an alleged criminal violation or other complex issue, safety and health professionals often forego the informal conference and immediately contest the citation and proceed to the Occupational Safety and Health Review Commission. Informal conferences can also be utilized as a fact-finding vehicle through which to possibly identify evidence OSHA has collected that may not be known to the safety professional. Lastly, the informal conference can be utilized to obtain specific answers and offer possible long-term solutions to difficult or challenging situations.

So what exactly is an informal conference? It is precisely as titled. In essence, an informal conference provides an opportunity for the safety professional and the area director for OSHA to sit down and talk outside of a formal hearing setting about the citations, proposed penalties, and virtually anything about the inspection, citations, or abatement. Informal conferences are often held in the conference room at the area OSHA office or other similar locations. The area director often asks the compliance officer who conducted the inspection and other OSHA officials to also attend, depending on the issues involved in the citation. The safety and health professional, members of the management team, legal counsel, and employees can participate in the informal discussions and provide documents, witness statements, or any support for their arguments or position. Legal counsel is not required, and employers usually do not request that legal counsel attend an informal conference. Employee representatives (usually union representatives) have the right to also participate in the informal conference.

Safety and health professionals should be aware that they must initiate the informal conference request. An informal conference is not automatic. The safety and health professional must contact the area OSHA office (usually on the top of the citation letterhead) and request an informal conference. The OSHA office will usually ask when the citation was issued and calculate the 15-workday limitation. Safety professionals must remember that the informal conference must be conducted prior to the deadline of 15 working days for contest. The OSHA office usually offers several available dates and times with the location usually at the area OSHA office.

It is important that the safety and health professional (after acquiring internal approvals) contact the area OSHA office and schedule the informal conference as soon as possible following the receipt of the citation. Remember, the informal conference must be scheduled during the 15-workday period and prior to the filing of the Notice of Contest. Safety and health professionals should be aware that if the matter is not settled at informal conference and the Notice of Contest is not filed before the 15-working-day limits, the employer or organization will lose all rights to contest or appeal the citations. Thus, time is of the essence for safety and health professionals to prepare for the informal conference, as well as preparing to protect the company or organization's rights to contest if the informal contest does not yield an acceptable settlement.

Preparation is the key to any successful informal conference. The first step in many companies and organizations is often the acquisition of the necessary approvals and authority to proceed to informal conference with OSHA and acquire the perimeters under which the safety professional may negotiate and settle each and every citation. In many larger companies and organizations, this may require substantial explanation of the citation, proposed penalties, abatement issues, and possible settlement perimeters at multiple levels within the organization in order to acquire the necessary authority to enter into settlement negotiations. Safety and health professionals must be well versed in all aspects of the citation in order to be able to fully explain the situation to the management team and others within the organizational structure.

Safety and health professionals should read and reread the citation acquiring a thorough knowledge of the citation document. The alleged violations identified in the citation should be verified and compared to the notes, photographs, videotape, or other documentation acquired during the inspection and as identified in the closing conference. Safety and health professionals should be aware that OSHA normally does not provide a copy of its files to assist the safety professional in preparing for an informal conference. Thus, it is vitally important that all notes, photographs, videotapes, and other documentation acquired before, during, and after the inspection are detailed and of the highest quality in order to assist the safety professional in preparing his or her arguments, positions, defenses, and possible solutions at the informal conference. In the event that the inspection was not documented, the safety professional is placed in the adverse position of attempting to re-create the inspection activities in order to acquire the information that can potentially be used at the informal conference. The risk, however, is that the information or situation documented by the compliance officer has changed or has been corrected by the time the safety and health professional does this. This can place the safety professional in an awkward position when OSHA provides substantial documentation supporting its position at the informal conference. Again, it is vitally important that each and every aspect of the inspection be documented in order to provide the necessary "ammunition" for the safety and health professional to prepare properly for an informal conference.

Safety and health professionals should evaluate the factors in establishing the penalties as identified in the OSHA *Field Operations Manual*[*] as well as in the recent OSHA *Administrative Penalty Information Bulletin*[†] to ascertain the applicability of these factors to the situations involved in the alleged violations. The first course of action for safety professionals is often an immediate repair or modification to the equipment, procedures, or other aspects identified as being deficient in the alleged violation and initiation of immediate corrective action. This corrective action should be well documented, including photographs, to be able to prove at the informal conference that the corrective actions have been taken and completed. The immediate repair or correction of the deficiency noted in the alleged violation reflects the safety professional's and the company's good faith and is grounds to request a reduction in degree categorization or proposed monetary penalty. The utilization of a good faith argument may not be applicable where there is a company history increase, a repeat violation, or the Severe Violator Enforcement Program is applicable.

Depending on the facts of the alleged violation, the gravity or seriousness of the alleged violation may be challenged if the safety and health professional can acquire and provide supporting evidence of a lower exposure or lower number of employees exposed to the alleged hazard. If applicable and supported by appropriate documentation, the safety and health professional may be able to argue that the circumstance involved in the alleged violation was an isolated incident creating increased seriousness, or the incident was created or controlled by a contractor or other entity.

Safety and health professionals should remember that their safety and health goals correlate with OSHA's goals to create and maintain a safe workplace for all employees. Although you may be working to achieve your goals from a different direction than OSHA, a safety and health professional, in essence, has the same or very similar goal as OSHA. And also remember, OSHA does not get to keep any of the monies acquired through monetary penalties. All paid monetary penalties go to the general fund. The monetary penalties are simply an enforcement method to penalize companies and organizations that have not maintained compliance with the laws and regulations. In essence, this is the "stick" to ensure all employers comply with the regulations governing the American workplace.

The safety and health professional should *always* be professional and open to discussion regarding any alleged violation or issues involved in the citation. An informal conference is the time to discuss openly in an attempt to resolve the citation. A safety and health professional should *never* take the alleged violations personally and be willing and open to hear OSHA's position on the alleged violations. The safety professional must be willing to listen intently and argue vigorously his or her position. Documentation often speaks louder than words. A prepared safety and health professional will be able to support his or her arguments or position with appropriate documentation.

---

[*] OSHA website, located at www.OSHA.gov.
[†] OSHA website, June 8, 2010.

When preparing for an informal conference, safety and health professionals should not only focus on the specific issues and defenses identified in the citation, but also step back and view the alleged violations on a broader scale from OSHA's perspective. When issuing a citation, OSHA possesses the burden of proving each and every element of each and every alleged hazard before the Occupational Safety and Health Review Commission and in a court of law. Safety and health professionals may want to examine each of the alleged violations within the citation from OSHA's perspective to explore and identify the supporting evidence that OSHA will utilize to meet its burden of proof. Any identified potential weaknesses in OSHA's position may be an avenue for discussion or negotiations at the informal conference. In the event that the safety professional can identify weaknesses in OSHA's position, there is often the potential that the OSHA area director recognizes the weakness in the case and thus may be willing to reduce the categorization or degree, and thus the monetary penalties, at an informal conference. However, with this being said, the OSHA compliance officers are well educated and experienced. Major deficiencies and weaknesses in OSHA's case are usually few and far between.

Again, during the preparation phase for the informal conference, safety and health professionals should not lose sight of the deadline to file the Notice of Contest in the event that the informal conference is not successful in settling the entire citation. Often safety and health professionals prepare their Notice of Contest in advance of the informal conference in order to be prepared to immediately file the Notice of Contest if the informal conference is not successful. Time is always of the essence once the 15-working-day clock starts ticking with the receipt of the citation. Safety and health professionals must be aware that if the citation is not settled completely at the informal conference or the Notice of Contest is not filed in a timely manner, the citation will become a final order and is usually not subject to review by the Occupational Safety and Health Review Commission, OSHA, or any other agency or court of law.

Safety and health professionals should consider the utilization of an informal conference for alleged violations and proposed monetary penalties. The cost to the company and organization is relatively small in comparison to a formal appeal, and many of these alleged violations and proposed penalties can be settled at the informal conference to the benefit of the company as well as OSHA. However, it is up to the safety and health professional to "move the ball" to request an informal conference and prepare, in a relatively short period of time, to address the alleged violations and proposed penalty. And remember, if you are not able to settle the citations at the informal conference, don't miss the 15-working-day deadline to appeal.

DISCUSSION QUESTIONS

1. When would it be beneficial to request an informal conference? Please explain in detail.
2. What happens if the company or organization does not appeal the citation(s) within 15 working days? Please explain in detail.
3. Where can the rules for the OSHRC be found?
4. Please explain the process through which a safety and health professional should prepare for an informal conference.

15 Future of the Safety and Health Profession

The only limit to our realization of tomorrow will be our doubts of today.

—Franklin D. Roosevelt

You can't escape the responsibility of tomorrow by evading it today.

—Abraham Lincoln

LEARNING OBJECTIVES

1. Analyze and assess the future of the safety and health function.
2. Analyze and assess the impact of the graying of the safety and health profession.
3. Analyze and assess the ethical challenges facing the safety and health profession.
4. Analyze and assess the impact of prescription and controlled substances on the American workplace and the safety and health profession.

Looking into the crystal ball, the future of the safety and health profession looks bright but potentially very different. First, the Occupational Safety and Health Act of 1970[*] is now a mature law and integrated into most organizations in the United States. Compliance, once the mainstay of the function of the safety and health professional, is still prominent, but a smaller percentage of the overall risk-related responsibilities of the safety and health professional. Second, the workplace has and will continue to change, becoming even more global in nature, more technical and automated, and required skills will be transformed from manual to specialized skills. Third, the workplace will continue to become more diverse, permitting all to pursue gainful employment. Fourth, the competition has and will continue to become globalized, with companies continuing to pursue a competitive advantage through technology, labor costs, transportation costs, and related functions. Lastly, the American workforce, as well as many others around the globe, is aging, creating a new dynamic within the labor force.

In the American workforce, the safety and health profession is graying as retirement looms for many baby boomers, thus creating new opportunities for younger professionals. However, given the relatively low number of students entering the safety and health profession via collegiate programs, competition among and

---

[*] Pub. L. 91-596 (1970).

between companies and organizations for the "best and brightest" will be stiff. However, given the recent economic downturn and the evaporation of many safety and health professionals' retirement "nest eggs" in the stock markets, the number of current safety and health professionals remaining in the workforce post scheduled retirement appears to have increased. Safety and health professionals, whether new to the profession or retooling for continued employment, must possess a significantly enhanced skill set to address issues and challenges not previously addressed.

As noted in previous chapters, safety and health professionals do not work in a vacuum and are today challenged with a multitude of new laws and regulations. Safety and health professionals today and in the future will be challenged with new and different issues addressing individual and worker rights under such laws as the Americans with Disabilities Act* and the Americans with Disabilities Amendments Act.† The Occupational Safety and Health Administration (OSHA) standards, which lie at the heart of the safety and health profession, have and will continue to develop new, more complex, and performance-challenging standards. With the changes in the American workplace, safety and health professionals will be challenged to appropriately train and educate employees ranging from 18 to 80 years of age and beyond.

Given the changing American and global workplace, the education and qualification levels for those entrusted with the safety and health of the employees should be appropriately qualified, tested, and regulated. Given the complexity and challenges ahead, companies and organizations can no longer promote a line worker or first-line supervisors into the positions responsible for the safety and health functions providing little or no actual training and education. The safety and health professional of the near future must be appropriately educated, qualified, tested, and licensed to ensure that he or she possesses the appropriate skill level to adequately manage the safety and health function in order to safeguard all employees. The day of the safety manager walking around with a clipboard and telling employees to put on their safety glasses is gone. Companies and organizations require a highly educated and skilled professional possessing all of the necessary abilities to effectively and efficiently manage the safety and health function as well as all correlating and ancillary functions. OSHA, state plan states, and safety and health professional organizations will be carefully assessing and analyzing the necessary qualifications of individuals who wish to work within the safety and health profession, and professional licensure, not unlike the legal and medical professions, is on the horizon.

Safety and health professionals will face new and challenging issues in the areas of ethics and professional conduct. Given the current lack of licensure and a mandatory code of conduct in the safety and health profession, current safety and health professionals are often left to flounder with little guidance or direction in addressing ethical challenges. Safety and health professionals currently encounter myriad professional conduct and ethical issues; however, it is anticipated that the sheer number of issues and challenges requiring guidance for and fortitude by the safety and health professional will increase, the applicable or correlating laws and regulations will increase, and the workplace will transform to adjust to the economy, legislation,

---

* Pub. L. 101-336 (1990).
† Pub. L. 102-166 (1991).

and competition. Safety and health professionals can be placed in ethical dilemmas where the safety and health professional is between "a rock and a hard place" with little or no guidance. However, the safety and health professional will most definitely be "second guessed" as to any decision and potentially could incur negative ramifications from the employer, the courts, or others.

Safety and health professionals are prone to the same addictions as others within our society. One major abuse issue is the use of prescription medication as well as the use of illegal controlled substances within our society, and arguably within a percentage of our safety and health professional population. The safety and health profession, unlike the legal and medical professions, possesses no ability or mechanism through which to provide rehabilitation assistance to safety and health professionals. Currently, safety and health professionals are "lone rangers" with no direction or assistance from the profession itself. In the future and as part of any licensure process, the safety and health profession, rather than individual employers, should provide assistance to individual safety and health professionals in addressing addiction issues.

In summation, the future within the occupational safety and health profession is froth with opportunity for those willing to accept the challenges. The safety and health profession will change as will the global workplace, laws and regulations, and workforce. The responsibilities of the safety and health professional will expand to encompass such other risk-related responsibilities as environment compliance, workplace security, and related areas. The employment future looks bright; however, the profession in and of itself may continue to be bifurcated while such issues as testing, licensure, and enforcement are negotiated and implemented. The safety and health profession is one of the few that have a direct impact on the lives and livelihoods of each and every employee in the workforce. Safety and health professionals will continue to impact the lives of many individuals, although through different routes, far into the future.

DISCUSSION QUESTIONS

1. Identify one potential impact area on the safety and health function and explain in detail.
2. What is the impact of prescription medication and controlled substances on the safety and health profession?
3. What is the impact of the global economy on the safety and health profession?
4. What is the safety and health profession going to look like in 10 years? 20 years?

Appendix A: Analyzing and Briefing a Court Decision

Safety and health professionals work with and for the law on a daily basis. In the United States, laws can be made by the executive branch (i.e., executive order) or the legislative branch (i.e., new laws), and the judicial branch is to review and assess the validity and applicability of these laws. This assessment by the judicial branch of our government is often called case law and is the evaluation, assessment, and decisions of the courts at all levels up to and including the U.S. Supreme Court.

Safety and health professionals should be aware that there are state courts as well as federal courts, and even specialty courts such as family law courts and traffic courts. Although the court name may vary, both the state and federal courts possess a hierarchy wherein each court decision can be appealed to a higher court by any of the parties involved in the actions. The highest court in most state judiciary systems as well as the federal judiciary system is the Supreme Court. For safety and health professionals who may be unfamiliar with their individual state judiciary system or the federal judiciary system, it is important to acquire a basic knowledge of the levels of the courts, namely, the specialty courts, such as tax court, the district or trial courts, where most trials take place, the appellate courts, and the state top court or the U.S. Supreme Court.

When reading a court decision, safety and health professionals may wish to utilize the following outline:

1. Identify the court and date of the case at the top of the decision.
2. Look up any legal terms you are not familiar with.
3. Read the case in total.
4. Determine the type of case you are reading.
5. Review the case summary or headnotes.
6. Read the case again and identify the court's decision.
7. If an appellate case, identify if the decision was unanimous or split.
8. Did the minority provide a dissenting opinion?
9. Reread the case. Identify the parties, issues, and facts of the case.
10. Brief your case in writing so you will remember the issues, facts, and decision at a later date.

Safety and health professionals should be aware that although the courts are not supposed to make law, their decisions are in fact shaping and making new law. The decisions of the courts are often called case law, which is the accumulation of court decisions that provide guidance and direction on current and future cases and decisions. As identified above, cases usually start at the lowest level in either the state or federal judiciary system and are appealed upward within the system to the top court, but can stop at any level. The decision of the higher court usually supersedes the decision of the lower court in whole or in part.

It is important that safety and health professionals acquire the skill and ability to carefully analyze these court decisions in order to know the status of the law on any given day. Safety professionals can find the court's decisions at most courthouses or law libraries; however, databases such as Westlaw and Lexis provide all cases to the safety and health professional as near as his or her computer. Court decisions are identified by name, *Jones v. Smith*, as well as the volume and page number within the identified location of the case and the year of the decision. As an example, *Jones v. Smith*, 22 U.S. 25 (2011). The parties are Jones and Smith, whom the action is against. Jones would be the plaintiff and Smith the defendant. The case can be counted in Volume 22 within the U.S. Supreme Court cases at page 25. The decision was rendered by the U.S. Supreme Court in 2011.

As identified above, it is important that the safety and health professional first read the case in full to identify the type of case, e.g., criminal or civil, as well as acquiring a flavor of the case. Safety and health professionals should identify the parties, the type of case, the broad issues, and the defenses as well as the court's decision. After the initial reading of the case, safety and health professionals should reread the case with an eye to the detail provided within it. On the third reading, the safety and health professional should takes notes and begin to assemble the structure of the case brief, which the safety professional can use to refresh his or her memory of the case or for use in whatever activity is at hand.

The primary reason safety and health professionals should brief a case is to provide an understanding of the particular issues and decisions in the cases as well as to provide a method of remembering the cases when a large number of cases are involved in the situation. Although there are various methods of briefing a case, the following is provided as one example. Safety professionals should find a method with which they are comfortable and utilize it consistently in their work.

BRIEFING A CASE METHODOLOGY

1. Case: List the case name, the court, and the date of the case at the top of the brief.
2. Issue: In a clear and concise manner of no more than one to three sentences, completely explain the issue(s) in the case.
3. Facts: In a clear and concise manner of no more than one or two paragraphs, identify all of the pertinent facts of the case.
4. Holding or decision: Clearly and concisely identify the decision of the court.
5. Dissent or dissenting opinion: If the minority provided a dissenting opinion, provide a clear and concise explanation of the dissenting judge's position.
6. Your opinion: It is important for safety professionals to identify why they agree or disagree with the decision of the court.

Most case briefs should be one page in length but no more than two pages. Safety and health professionals who exceed the one-page limit may want to reassess their analysis and reduce the verbiage to address only the issues, facts, and decision provided by the court. In essence, the brief is a short, concise, and to-the-point document that safety professionals can use to remember the case and possess a quick review of the important aspects of the case. If additional details are needed, the safety professional should go back and review the entire case.

Below is an example of a very basic case brief for your review and use:

Case Name: *Smith v. Jones, Inc.*, 10 Any state Court, 21 (2011)

Issues: Smith alleged she was discriminated against by her employer, Jones, Inc., in violation of the Pregnancy Discrimination Act and Title VII of the Civil Rights Act.

Facts: Smith, employed by Jones, Inc. for 5 years, was terminated from her employment. Jones alleges Smith was terminated for theft of company property. Smith alleges she was terminated when she informed her boss that she was pregnant.

Holding: For the defendant, Jones, Inc. The court found that the termination of Smith for theft was appropriate and found no discrimination based on her pregnancy.

My opinion: I disagree with the decision of the court. Jones possessed a history of terminating employees who filed for any type of leave of absence. Smith was constructively terminated under the precursor of theft, which was simply a company pen she forgot in her purse.

Appendix B: Surfing the OSHA Website

A great source of information for safety and health professionals is the Occupational Safety and Health Administration (OSHA) website located at www.osha.gov. This governmental website possesses all of the standards as well as interpretations, statistics, data, training tools, and latest news releases. This website has quickly become the "go to" source for many safety and health professionals.

The OSHA website is very user-friendly and possesses an excellent search process. The website also possesses links to other correlating websites, such as the National Institute for Occupational Safety and Health (NIOSH), and links to related publications.

Of particular usefulness for safety and health professionals are the standards, field manual, and statistics. Each and every standard in general industry as well as construction can be easily located on the site or through a search. The interpretations for each of these standards are also easily searched by switching to the interpretation search database. The website also contains different statistical databases, including the Bureau of Labor statistics, and the most current injury and fatality statistics.

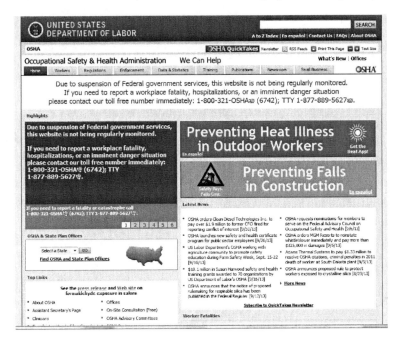

Appendix C: Developing a Compliance Program

One of the most important tasks for a safety and health professional is the development of a written compliance program. The written compliance program provides guidance to the management and employees as well as serves as "proof" for the purposes of compliance. So, where do you start?

1. Evaluate and analyze the standard to ensure its applicability to your situation.
2. Locate the standard in the Code of Federal Regulations (CFR) or on the Occupational Safety and Health Administration (OSHA) website: www.osha.gov.
3. Read the standard. Reread the standard.
4. Outline or identify the key points and requirements in the standard.
5. Apply the standard to your workplace.
6. Visualize your program in action.
7. Draft your written program.
8. Ensure *all* elements of the standard are addressed. Insert operational elements necessary for implementation in your workplace.
9. Review your program and have your program reviewed within your corporate structure.
10. Distribute, acquire necessary internal approvals, and implement your program.

Below please find several program examples from the OSHA website (www.osha.gov). I would caution students to only use these samples as a framework for your original compliance program, which is usually unique to your operations. And remember to include other such elements as necessary to implement or manage your program, such a disciplinary section for enforcement of your program.

SAMPLE PROGRAMS

The following sample safety and health programs are intended to provide examples of written programs on various workplace safety and health topics. They are not intended to supersede the requirements in OSHA standards. Employers should consult the applicable OSHA standards for the specific requirements applicable to their workplaces. Employers can use these sample programs as guidance when developing their own customized programs that are tailored to their specific workplaces.

- Bloodborne Pathogens
- Fall Protection
- General Safety and Health Programs
- Hazard Communication Standard

- Lockout/Tagout
- Powered Industrial Trucks
- Respiratory Protection
- Shipyard Employment
- Tuberculosis
- OSHA eTools for Developing Customized Programs
- Sample Programs from State On-Site Consultation Programs

BLOODBORNE PATHOGENS

- *Model Plans and Programs for the OSHA Bloodborne Pathogens and Hazard Communications Standards.* OSHA Publication 3186. Also available as a 520 KB PDF.

FALL PROTECTION

- Sample Fall Protection Plan for Residential Construction. Also available as a 43 KB DOC.
- Sample Fall Protection Plans for Construction. Appendix E to 29 CFR 1926, Subpart M.

GENERAL SAFETY AND HEALTH PROGRAMS

- Sample Safety and Health Program for Small Business. Included in the OSHA Safety and Health Management Systems eTool.
- Sample Safety and Health Programs for the Logging Industry. Part of OSHA's Logging Advisor eTool.

HAZARD COMMUNICATION STANDARD

- *Model Plans and Programs for the OSHA Bloodborne Pathogens and Hazard Communications Standards.* OSHA Publication 3186. Also available as a 521 KB PDF.

LOCKOUT/TAGOUT

- Typical Minimal Lockout Procedures. Appendix A to 29 CFR 1910.147.

POWERED INDUSTRIAL TRUCKS

- Outline of Sample Operator Training Program. Also available as a 12 KB PDF. Included in training materials developed by OSHA directorate of training and education.

Respiratory Protection

- Sample Respiratory Protection Program. Attachment 4, page 103, Appendix IV in *Small Entity Compliance Guide for the Revised Respiratory Protection Standard.*

Shipyard Employment

- Model Fire Safety Plan. Appendix A to 29 CFR 1915, Subpart P.

Tuberculosis

- Sample Exposure Control Plan. Part of OSHA's eTools for hospitals and nursing homes. This sample plan was included in OSHA's proposed tuberculosis standard (Appendix F to proposed 29 CFR 1910.1035, 62 Fed. Reg. 54,160, October 17, 1997), but OSHA withdrew the proposed rule in May 2003.

OSHA eTools for Developing Customized Programs

- Confined Spaces: An OSHA eTool (public test version) helps employers prepare a written permit-required confined space program.
- Emergency Action Plan: An OSHA eTool helps employers create their own basic emergency action plan.
- Fire Prevention Plan: An OSHA eTool helps employers write customized emergency action plans and fire prevention plans.
- Hazardous Waste Site Operations: An OSHA/EPA eTool (e-HASP) helps employers develop written site-specific health and safety plans.
- Lockout/Tagout: OSHA's Lockout/Tagout Interactive Training Program includes guidance on developing an energy control program.

Sample Programs from State On-Site Consultation Programs

A number of on-site consultation programs provide collections of sample safety and health programs on their websites, including those listed below. The on-site consultation programs are run by the states with funding from OSHA. If you are in a state with an OSHA approved state program, please check with your state agency.

These sample programs provide examples of written programs on various workplace safety and health topics. They are not intended to supersede the requirements in OSHA standards. Employers should consult the applicable OSHA standards when developing their own customized programs that are tailored to their workplace. Contact your state on-site consultation program to request help in developing customized programs.

- Alabama
- Arkansas

- Connecticut
- Idaho: General Industry and Construction
- Illinois
- Massachusetts
- Missouri
- Montana
- Nebraska
- Texas*

---

* OSHA website: www.osha.gov.

Appendix D: Citation Document Taken from the OSHA Website[*]

---

[*] This citation document has been edited for the purposes of this text.

U.S. Department of Labor
Occupational Safety and Health Administration
Shattuck Office Center
138 River Road, Suite 102
Andover, MA 01810
Phone: (978)837-4460 FAX: (978) 837-4455

Citation and Notification of Penalty

To:

Inspection Number:	▰▰▰▰▰
Inspection Date(s):	03/13/2011-03/16/2011
Issuance Date:	09/12/2011

Inspection Site:

The violation(s) described in this Citation and Notification of Penalty is (are) alleged to have occurred on or about the day(s) the inspection was made unless otherwise indicated within the description given below.

This Citation and Notification of Penalty (this Citation) describes violations of the Occupational Safety and Health Act of 1970. The penalty(ies) listed herein is (are) based on these violations. You must abate the violations referred to in this Citation by the dates listed and pay the penalties proposed, unless within 15 working days (excluding weekends and Federal holidays) from your receipt of this Citation and Notification of Penalty you mail a notice of contest to the U.S. Department of Labor Area Office at the address shown above. Please refer to the enclosed booklet (OSHA 3000) which outlines your rights and responsibilities and which should be read in conjunction with this form. Issuance of this Citation does not constitute a finding that a violation of the Act has occurred unless there is a failure to contest as provided for in the Act or, if contested, unless this Citation is affirmed by the Review Commission or a court.

Posting - The law requires that a copy of this Citation and Notification of Penalty be posted immediately in a prominent place at or near the location of the violation(s) cited herein, or , if it is not practicable because of the nature of the employer's operations, where it will be readily observable by all affected employees. This Citation must remain posted until the violation(s) cited herein has (have) been abated, or for 3 working days (excluding weekends and Federal holidays), whichever is longer. **The penalty dollar amounts need not be posted and may be marked out or covered up prior to posting.**

Informal Conference - An informal conference is not required. However, if you wish to have such a conference you may request one with the Area Director during the 15 working day contest period. During such an informal conference you may present any evidence or views which you believe would support an adjustment to the citation(s) and/or penalty(ies).

If you are considering a request for an informal conference to discuss any issues related to this Citation and Notification of Penalty, you must take care to schedule it early enough to allow time to contest after the informal

---

conference, should you decide to do so. Please keep in mind that a written letter of intent to contest must be submitted to the Area Director within 15 working days of your receipt of this Citation. The running of this contest period is not interrupted by an informal conference.

If you decide to request an informal conference, please complete, remove and post the page 4 Notice to Employees next to this Citation and Notification of Penalty as soon as the time, date, and place of the informal conference have been determined. Be sure to bring to the conference any and all supporting documentation of existing conditions as well as any abatement steps taken thus far. If conditions warrant, we can enter into an informal settlement agreement which amicably resolves this matter without litigation or contest.

Right to Contest - You have the right to contest this Citation and Notification of Penalty. You may contest all citation items or only individual items. You may also contest proposed penalties and/or abatement dates without contesting the underlying violations. **Unless you inform the Area Director in writing that you intend to contest the citation(s) and/or proposed penalty(ies) within 15 working days after receipt, the citation(s) and the proposed penalty(ies) will become a final order of the Occupational Safety and Health Review Commission and may not be reviewed by any court or agency.**

Penalty Payment - Penalties are due within 15 working days of receipt of this notification unless contested. (See the enclosed booklet and the additional information provided related to the Debt Collection Act of 1982.) Make your check or money order payable to "DOL-OSHA". Please indicate the Inspection Number on the remittance.

OSHA does not agree to any restrictions or conditions or endorsements put on any check or money order for less than the full amount due, and will cash the check or money order as if these restrictions, conditions, or endorsements do not exist.

Notification of Corrective Action - For **each** violation which you do not contest, you are required by 29 CFR 1903.19 to submit an Abatement Certification to the Area Director of the OSHA office issuing the citation and identified above. The certification **must** be sent by you within **10 calendar days** of the abatement date indicated on the citation. For **Willful** and **Repeat** violations, documents (examples: photos, copies of receipts, training records, etc.) demonstrating that abatement is complete must accompany the certification. Where the citation is classified as **Serious** and the citations states that abatement documentation is required, documents such as those described above are required to be submitted along with the abatement certificate. If the citation indicates that the violation was corrected during the inspection, no abatement certification is required for that item.

All abatement verification documents must contain the following information: 1) Your name and address; **2)** the inspection number (found on the front page); **3)** the citation and citation item number(s) to which the submission relates; **4)** a statement that the information is accurate; **5)** the signature of the employer or employer's authorized representative; **6)** the date the hazard was corrected; **7)** a brief statement of how the hazard was corrected; and **8)** a statement that affected employees and their representatives have been informed of the abatement.

The law also requires a copy of all abatement verification documents, required by 29 CFR 1903.19 to be sent to OSHA, also be posted at the location where the violation appeared and the corrective action took place.

Employer Discrimination Unlawful - The law prohibits discrimination by an employer against an employee for filing a complaint or for exercising any rights under this Act. An employee who believes that he/she has been discriminated against may file a complaint no later than 30 days after the discrimination occurred with the U.S. Department of Labor Area Office at the address shown above.

Employer Rights and Responsibilities - The enclosed booklet (OSHA 3000) outlines additional employer rights and responsibilities and should be read in conjunction with this notification.

Notice to Employees - The law gives an employee or his/her representative the opportunity to object to any abatement date set for a violation if he/she believes the date to be unreasonable. The contest must be mailed to the U.S. Department of Labor Area Office at the address shown above and postmarked within 15 working days (excluding weekends and Federal holidays) of the receipt by the employer of this Citation and Notification of penalty.

Inspection Activity Data - You should be aware that OSHA publishes information on its inspection and citation activity on the Internet under the provisions of the **Electronic Freedom of Information Act.** The information related to your inspection will be available 30 calendar days after the Citation Issuance Date. You are encouraged to review the information concerning your establishment at WWW.OSHA. GOV. If you have any dispute with the accuracy of the information displayed, please contact this office.

U.S. Department of Labor
Occupational Safety and Health Administration

NOTICE TO EMPLOYEES OF INFORMAL CONFERENCE

An informal conference has been scheduled with OSHA to discuss the citation(s) issued on

09/12/2011. The conference will be held at the OSHA office located at ▆▆▆▆▆▆▆▆▆▆▆▆▆▆,

▆▆▆▆▆▆▆▆▆▆▆▆▆▆▆▆▆▆▆▆▆▆▆ on _____ at _____.

Employees and/or representatives of employees have a right to attend an informal conference.

U.S. DEPARTMENT OF LABOR

OCCUPATIONAL SAFETY AND HEALTH ADMINISTRATION

CERTIFICATE OF CORRECTION

The undersigned certifies that on _____ , all of the violations cited on
                                            (date)

OSHA Citation # _____ issued on _____ , were corrected and that a
                                                    (date)

copy of this Certificate of Correction was posted on _____ in a manner and place
                                                            (date)

for review by affected employees.

_____
Employer's Signature

U.S. Department of Labor
Occupational Safety and Health Administration

Inspection Number:
Inspection Dates: 03/13/2011 - 03/16/2011
Issuance Date: 09/12/2011

Citation and Notification of Penalty

Company Name: ▮▮▮▮▮▮
Inspection Site: ▮▮▮▮▮▮▮▮▮▮▮▮▮

Citation 1 Item 1   Type of Violation: **Serious**

29 CFR 1910.37(a)(1) Exit routes were not kept free of explosive or highly flammable furnishings or other decorations:

   (a) Location - Building 24:
   On or about 3/13/11, numerous fifty five gallon drums of flammable liquids were stored along the exit access to the south west exit.

   (b) Location - Building 25:
   On or about 3/13/11, two fifty five gallon drums of flammable liquids were stored in the exit access of the specialty area in the southwest corner of the building.

Date By Which Violation Must be Abated:	10/27/2011
Proposed Penalty:	$ 7000.00

Citation 1 Item 2   Type of Violation: **Serious**

29 CFR 1910.120(q)(1):An emergency response plan shall be developed and implemented to handle anticipated emergencies prior to the commencement of emergency response operations. Employers who will evacuate their employees from the danger area when an emergency occurs, and who do not permit any of their employees to assist in handling the emergency, are exempt from the requirements of this paragraph if they provide an emergency action plan in accordance with 1910.38(a):

   (a) Location - ▮▮▮▮▮▮▮▮▮▮▮
   On or about 3/13/11, the employer did not implement an adequate emergency response plan to handle anticipated emergencies prior to commencement of emergency response operations.

Date By Which Violation Must be Abated:	10/27/2011
Proposed Penalty:	$ 7000.00

See pages 1 through 4 of this Citation and Notification of Penalty for information on employer and employee rights and responsibilities.

U.S. Department of Labor
Occupational Safety and Health Administration

Inspection Number:
Inspection Dates: 03/13/2011-03/16/2011
Issuance Date: 09/12/2011

Citation and Notification of Penalty

Company Name:
Inspection Site:

<u>Citation 1 Item 3</u> Type of Violation: **Serious**

29 CFR 1910.120 (q)(2): The employer shall develop an emergency response plan for emergencies which shall address, as a minimum, elements described in 1910.120(q)(i) through (xii):

   (a) Location -
   On or about 3/13/11, the employer's emergency response plan did not adequately nor effectively address elements included in 1910.120(q)(2) including but not limited to the following:
   - (ii) Personnel roles, lines of authority, training, and communication,
   - (iii) Emergency recognition and prevention,
   - (iv) Safe distances and places of refuge,
   - (vi) Evacuation routes and procedures,
   - (vii) Decontamination,
   - (ix) Emergency alerting and response procedures,
   - (x) Critique of response and follow-up,
   - (xi) Personal protective equipment and emergency equipment.

Date By Which Violation Must be Abated: 10/27/2011
Proposed Penalty: $ 7000.00

See pages 1 through 4 of this Citation and Notification of Penalty for information on employer and employee rights and responsibilities.

U.S. Department of Labor
Occupational Safety and Health Administration

Inspection Number:
Inspection Dates: 03/13/2011 - 03/16/2011
Issuance Date: 09/12/2011

Citation and Notification of Penalty

Company Name:
Inspection Site:

The alleged violations below have been grouped because they involve
similar or related hazards that may increase the potential for injury
resulting from an accident.

Citation 1 Item 4a Type of Violation: **Serious**

29 CFR 1910.120(q)(3)(i): The senior emergency response official responding to an emergency did not become the individual in charge of a site specific Incident Command System (ICS) and/or coordinate and control communications.

> (a) Location -
> On or about 3/13/11, the employer did not take adequate precautions to coordinate and/or control use of communications equipment including, but not limited to; portable radios, cell phones, and other equipment which were not rated for use in the presence of flammable liquids and/or vapors.

Date By Which Violation Must be Abated: 10/27/2011
Proposed Penalty: $ 7000.00

Citation 1 Item 4b Type of Violation: **Serious**

29 CFR 1910.106(h)(7)(i)(a): Precautions were not taken to prevent the ignition of flammable vapors:

> (a) Location -
> On or about 3/13/11, the employer did not take adequate precautions to prevent the ignition of flammable vapors when using radios, cell phones, and other equipment which were not intrinsically safe.

Date By Which Violation Must be Abated: 10/27/2011

See pages 1 through 4 of this Citation and Notification of Penalty for information on employer and employee rights and responsibilities.

U.S. Department of Labor
Occupational Safety and Health Administration

Inspection Number:
Inspection Dates: 03/13/2011 - 03/16/2011
Issuance Date: 09/12/2011

Citation and Notification of Penalty

Company Name:
Inspection Site:

<u>Citation 1 Item 5</u> Type of Violation: **Serious**

29 CFR 1910.120(q)(6)(iii): Employees who participated or were expected to participate, as hazardous materials technicians had not received at least 24 hours of training equal to the first responder operations level and/or had not been certified as having such training and/or in addition did not have competency in (A) through (I) of this section:

   (a) Location -
On or about 3/13/11, numerous employees that the employer's Integrated Contingency Plan (ICP) identified as being required to participate in emergencies as hazardous material technicians had not received at least 24 hours of training equal to the first responder operations level and in addition had not been certified as competent in (A) through (I) of this section.

Date By Which Violation Must be Abated:	10/27/2011
Proposed Penalty:	$ 7000.00

<u>Citation 1 Item 6</u> Type of Violation: **Serious**

29 CFR 1910.120(q)(6)(v): Employees who participated, or were expected to participate, as an on-scene incident commander had not received 24 hours of training equal to first responder operations level and/or had not been certified as having such training and/or in addition did not have the competency in areas described in 29 CFR 1910.120.(q)(6)(v)(A-F) and so certified:

   (a) Location -
On or about 3/13/11, employees expected to perform incident commander duties had not received twenty four hours required training at the first responder operations level and had not been certified as having competency in (A) through (F) of this section.

Date By Which Violation Must be Abated:	10/27/2011
Proposed Penalty:	$ 7000.00

See pages 1 through 4 of this Citation and Notification of Penalty for information on employer and employee rights and responsibilities.

U.S. Department of Labor
Occupational Safety and Health Administration

Inspection Number:
Inspection Dates: 03/13/2011 - 03/16/2011
Issuance Date: 09/12/2011

Citation and Notification of Penalty

Company Name:
Inspection Site:

<u>Citation 1 Item 7</u> Type of Violation: **Serious**

29 CFR 1910.120(q)(8)(i): Employees who were trained in accordance with 29 CFR 1910.120(q)(6) did not receive annual refresher training of sufficient content and duration to maintain their competencies or did not demonstrate competency in those areas at least yearly:

   (a) Location -
   On or about 3/13/11, the employer did not provide adequate and/or effective annual refresher training or a chance for employees to demonstrate continued competency in implementing the integrated contingency plan in the event of an emergency response to a release of hazardous chemicals.

   Date By Which Violation Must be Abated: 10/27/2011
   Proposed Penalty: $ 7000.00

<u>Citation 1 Item 8</u> Type of Violation: **Serious**

29 CFR 1910.120(q)(10): Chemical protective clothing and equipment to be used by organized and designated HAZMAT team members, or to be used by hazardous materials specialists, did not meet the requirements of 29 CFR 1910.120(g)(3) through (5):

   (a) Location -
   On or about 3/13/11, the employer did not establish a written personal protective equipment program for employees who are members of the on site Emergency Response Team (ERT).

   Date By Which Violation Must be Abated: 10/27/2011
   Proposed Penalty: $ 7000.00

See pages 1 through 4 of this Citation and Notification of Penalty for information on employer and employee rights and responsibilities.

U.S. Department of Labor
Occupational Safety and Health Administration

Inspection Number:
Inspection Dates: 03/13/2011 - 03/16/2011
Issuance Date: 09/12/2011

Citation and Notification of Penalty

Company Name:
Inspection Site:

Citation 1 Item 9 Type of Violation: **Serious**

29 CFR 1910.134(f)(1): The employer did not ensure that employee(s) using a tight-fitting face-piece respirator passed an appropriate qualitative fit test (QLFT) or quantitative fit test (QNFT):

(a) Location -
On or about 3/13/11, employees on the emergency response team were expected to use a tight-fitting face-piece respirator as part of a self-contained breathing apparatus (SCBA) and had not been fit tested.

Date By Which Violation Must be Abated:	10/27/2011
Proposed Penalty:	$ 7000.00

Citation 1 Item 10 Type of Violation: **Serious**

29 CFR 1910.157(g)(3): Employees who have been designated to use fire fighting equipment as part of the emergency action plan were not provided training in the use of the appropriate equipment:

(a) Location -
On or about 3/13/11, employees who are required to use fire extinguishers to extinguish fires of substantial quantities of flammable liquids were not provided with adequate training in the use of the appropriate equipment.

Date By Which Violation Must be Abated:	10/27/2011
Proposed Penalty:	$ 7000.00

See pages 1 through 4 of this Citation and Notification of Penalty for information on employer and employee rights and responsibilities.

Citation and Notification of Penalty   Page 11 of 19   OSHA-2 (Rev. 9/93)

U.S. Department of Labor
Occupational Safety and Health Administration

Inspection Number:
Inspection Dates: 03/13/2011 - 03/16/2011
Issuance Date: 09/12/2011

Citation and Notification of Penalty

Company Name: ▮▮▮▮▮▮▮▮
Inspection Site:

Citation 1 Item 11 Type of Violation: **Serious**

29 CFR 1910.157(g)(4): Employees who have been designated to use fire fighting equipment as a part of the emergency action plan were not provided training in the use of appropriate equipment upon initial assignment and at least annually thereafter:

    (a) Location - ▮▮▮▮▮▮▮▮
    On or about 3/13/11, employees required to use fire extinguishers were not provided training at least annually in the use of appropriate equipment.

    Date By Which Violation Must be Abated: 10/27/2011
    Proposed Penalty: $ 7000.00

Citation 1 Item 12 Type of Violation: **Serious**

29 CFR 1910.252(a)(2)(iv): Cutting or welding was permitted in areas before the areas were inspected by the individual responsible for authorizing cutting and welding operations:

    (a) Location - ▮▮▮▮▮▮▮▮
    On or about 3/13/11, the employer failed to ensure that areas were inspected prior to welding and cutting being permitted.

    Date By Which Violation Must be Abated: 10/27/2011
    Proposed Penalty: $ 7000.00

See pages 1 through 4 of this Citation and Notification of Penalty for information on employer and employee rights and responsibilities.

U.S. Department of Labor
Occupational Safety and Health Administration

Inspection Number:
Inspection Dates: 03/13/2011 - 03/16/2011
Issuance Date: 09/12/2011

Citation and Notification of Penalty

Company Name:
Inspection Site:

Citation 2 Item 1 Type of Violation: **Willful**

29 CFR 1910.307(c)(2)(i): Equipment shall be approved not only for the class of location, but also for the ignitable or combustible properties of the specific gas, vapor, dust, or fiber that will be present.

(a) Location -
On or about 3/13/11, the light fixture and operating switch in a hazardous location on top of the Solvator was not approved for Class I hazardous locations nor for the ignitable properties of the vapor.

Date By Which Violation Must be Abated:	10/27/2011
Proposed Penalty:	$ 70000.00

Citation 2 Item 2 Type of Violation: **Willful**

29 CFR 1910.307(c)(2)(i): Equipment shall be approved not only for the class of location, but also for the ignitable or combustible properties of the specific gas, vapor, dust, or fiber that will be present.

(a) Location -
On or about 3/13/11, the service disconnect switch, which was in a hazardous location, for the reflux pump adjacent to Reactor 8 was not approved for Class I hazardous locations nor for the ignitable properties of the vapor.

Date By Which Violation Must be Abated:	10/27/2011
Proposed Penalty:	$ 70000.00

See pages 1 through 4 of this Citation and Notification of Penalty for information on employer and employee rights and responsibilities.

U.S. Department of Labor
Occupational Safety and Health Administration

Inspection Number:
Inspection Dates: 03/13/2011 - 03/16/2011
Issuance Date: 09/12/2011

Citation and Notification of Penalty

Company Name:
Inspection Site:

<u>Citation 2 Item 3</u> Type of Violation: **Willful**

29 CFR 1910.307(c)(2)(i): Equipment shall be approved not only for the class of location, but also for the ignitable or combustible properties of the specific gas, vapor, dust, or fiber that will be present.

   (a) Location -
   On or about 3/13/11, the motor for Reactor 6 hot oil pump, which was in a hazardous location, was not approved for a Class I hazardous location nor for the ignitable properties of the vapor.

Date By Which Violation Must be Abated:	10/27/2011
Proposed Penalty:	$ 70000.00

<u>Citation 2 Item 4</u> Type of Violation: **Willful**

29 CFR 1910.307(c)(2)(i): Equipment shall be approved not only for the class of location, but also for the ignitable or combustible properties of the specific gas, vapor, dust, or fiber that will be present.

   (a) Location -
   On or about 3/13/11, the electrical timer switch and enclosure for Reactor 8's internal sight light, both of which were in a hazardous location, were not approved for Class I hazardous locations nor for the ignitable properties of the vapor.

Date By Which Violation Must be Abated:	10/27/2011
Proposed Penalty:	$ 70000.00

See pages 1 through 4 of this Citation and Notification of Penalty for information on employer and employee rights and responsibilities.

U.S. Department of Labor
Occupational Safety and Health Administration

Inspection Number:
Inspection Dates: 03/13/2011 - 03/16/2011
Issuance Date: 09/12/2011

Citation and Notification of Penalty

Company Name:
Inspection Site:

Citation 2 Item 5 Type of Violation: Willful

29 CFR 1910.307(c)(2)(i): Equipment shall be approved not only for the class of location, but also for the ignitable or combustible properties of the specific gas, vapor, dust, or fiber that will be present.

(a) Location -
On or about 3/13/11, the motor switch and enclosure for the exhaust fan for Reactor 8, both of which were in a hazardous location, were not approved for Class I hazardous locations nor for the ignitable properties of the vapor.

Date By Which Violation Must be Abated: 10/27/2011
Proposed Penalty: $ 70000.00

Area Director

See pages 1 through 4 of this Citation and Notification of Penalty for information on employer and employee rights and responsibilities.

U.S. Department of Labor
Occupational Safety and Health Administration
Shattuck Office Center
138 River Road, Suite 102
Andover, MA 01810
Phone: (978)837-4460 FAX: (978)837-4455

INVOICE/
DEBT COLLECTION NOTICE

Company Name:	▰▰▰▰▰
Inspection Site:	▰▰▰▰▰
Issuance Date:	09/12/2011

Summary of Penalties for Inspection Number ▰▰▰▰▰

Citation 1, Serious	=	**$ 126,000.00**
Citation 2, Willful	=	**$ 350,000.00**
TOTAL PROPOSED PENALTIES	=	**$ 476,000.00**

To avoid additional charges, please remit payment promptly to this Area Office for the total amount of the uncontested penalties summarized above. Make your check or money order payable to: "DOL-OSHA". Please indicate OSHA's Inspection Number (indicated above) on the remittance.

OSHA does not agree to any restrictions or conditions put on any check or money order for less than the full amount due and will cash the check or money order as if these restrictions or conditions do not exist.

If a personal check is issued, it will be converted into an electronic fund transfer (EFT). This means that our bank will copy your check and use the account information on it to electronically debit your account for the amount of the check. The debit from your account will then usually occur within 24 hours and will be shown on your regular account statement. You will not receive your original check back. The bank will destroy your original check, but will keep a copy of it. If the EFT cannot be completed because of insufficient funds or closed account, the bank will attempt to make the transfer up to 2 times.

Pursuant to the Debt Collection Act of 1982 (Public Law 97-365) and regulations of the U.S. Department of Labor (29 CFR Part 20), the Occupational Safety and Health Administration is required to assess interest, delinquent charges, and administrative costs for the collection of delinquent penalty debts for violations of the Occupational Safety and Health Act.

Interest. Interest charges will be assessed at an annual rate determined by the Secretary of the Treasury on all penalty debt amounts not paid within one month (30 calendar days) of the date on which the debt amount becomes due and payable (penalty due date). The current interest rate is 5%. Interest will accrue from the date on which the penalty amounts (as proposed or adjusted) become a final order of the Occupational Safety and Health Review Commission (that is, 15 working days from your receipt of the Citation and Notification of Penalty), unless you file a notice of contest. Interest charges will be waived if the full amount owed is paid within 30 calendar days of the final order.

Delinquent Charges. A debt is considered delinquent if it has not been paid within one month (30 calendar days) of the penalty due date or if a satisfactory payment arrangement has not been made. If the debt remains delinquent for more than 90 calendar days, a delinquent charge of six percent (5%) per annum will be assessed accruing from the date that the debt became delinquent.

Administrative Costs. Agencies of the Department of Labor are required to assess additional charges for the recovery of delinquent debts. These additional charges are administrative costs incurred by the Agency in its attempt to collect an unpaid debt. Administrative costs will be assessed for demand letters sent in an attempt to collect the unpaid debt.

Area Director

Date

Appendix E: Defenses to OSHA Alleged Violations

Upon receipt of a citation alleging violations and proposed penalties from the Occupational Safety and Health Administration (OSHA), the safety and health professional is often asked to provide the possible defenses and options to the management team. The safety and health professional may wish to explore any applicable and known defense that is appropriate to the facts and circumstances applicable to the alleged violation.

Procedural defenses include but are not limited to the following:

1. Statute of limitation defense: Generally, more than 6 months between the closing conference and issuance of citations.
2. Improper service defense: Citations not received in a timely manner.
3. State plan coverage defense: Jurisdiction by state plan rather than federal OSHA.
4. Lack of employment relationship defense: Not your employees.
5. Preempted by another federal agency defense.
6. The vagueness defense: The standard was vague on the particular issue.

Other procedural defenses may be available. Prudent safety and health professionals should consult their legal counsel.

Factual defenses include but are not limited to the following:

1. The isolated incident defense
2. The greater hazard to comply with the standard defense
3. The impossibility to comply with the standard defense
4. The lack of the employer's knowledge defense
5. The machine was not in use or energized defense

Other factual defenses may be available depending on the circumstances. Prudent safety and health professionals should consult their legal counsel.

Appendix F: What Is a Variance?

Variances

Overview

Variances from the Occupational Safety and Health Administration (OSHA) standards are authorized under sections 6 and 16 of the Occupational Safety and Health Act of 1970 (OSHAct), and the implementing rules contained in the Code of Federal Regulations (29 CFR 1905). A variance may be requested by an employer or by a class of employers for specific workplaces.

Requests for variance should be made as follows: for permanent variance [authorized by section 6(d) of the OSHAct] the requirements of 29 CFR 1905.11 are to be followed, while for temporary variance [authorized by section 6(b)(6)(A) of the OSHAct] the requirements of 29 CFR 1905.10 are applicable. There are also provisions in section 16 of the OSHAct (29 CFR 1905.12) for National Defense variances, and in section 6(b)(6)(C) of the OSHAct for variances allowing authorized experimentation. Each type variance is summarized below.

Permanent Variance

A permanent variance authorizes an alternative to a requirement of an OSHA standard as long as the applicant's employees are provided with employment and a place of employment equivalently safe and healthful. Two factors that must be addressed in the application are proof that the alternative is safe and healthful as the standard requirement, and that the employees have been appropriately notified of the request and of their rights.

The final determination by the Assistant Secretary for grant of permanent variance is based upon the employer's application and evidence, an on-site visit to the workplace by OSHA representatives, as deemed necessary, and comments by employees and other interested parties. If the request is granted, the final order details the differences between the requirements of the standard and the alternative, and specifies the employer's responsibilities and requirements.

Temporary Variance

A temporary variance is designed to provide an employer time to come into compliance with the requirements of an OSHA standard subsequent to the effective date of a standard. The request must be made within a reasonable time after the promulgation and prior to the effective date of the standard. This limited-time temporary order is based upon the employer demonstrating inability to comply with the standard by its effective date for one of the reasons specified in the OSHAct. The applicant must also establish that his employees are being adequately protected against the hazards covered by the standard, and that he has an effective program for coming into compliance with the standard as quickly as possible. The employer must also notify his employees of the request and of their rights.

The Assistant Secretary may issue a time-limited interim order pending the decision on the temporary variance, if such an order was requested by the applicant.

Experimental Variance

The Assistant Secretary is authorized to grant a variance whenever it is determined that such variance is necessary to permit an employer to participate in an experiment designed to demonstrate or validate new and improved techniques to protect the health or safety of workers.

Variations, Tolerances, and Exemptions

The Assistant Secretary, after notice and opportunity for a hearing, is permitted to grant reasonable variation, tolerance, or exception from the ACT's requirements in order to avoid serious impairment of the National Defense. Such grants shall not remain in effect for more than six months without further notifying affected employees and again affording the opportunity for a hearing.*

---

* From OSHA's website: www.osha.gov.

How to Apply for a Variance

OSHA has no single, uniform application form for an employer to apply for a variance. Therefore, to apply for a variance, an employer must review the specific regulations applicable to each type of variance, and submit the required information. Generally, the application can be in the form of a letter with the following information included:

- An explicit request for a variance

- The specific standard from which the employer is seeking the variance.

- Whether the employer is applying for a permanent, temporary, experimental, national defense, or recordkeeping variance, and an interim order. (If the application is for a temporary variance, state when the employer will be able to comply with the OSHA standard.)

- Describe the alternative means of compliance with the standard from which the applicant is seeking the variance. The statement must contain sufficient detail to support, by a preponderance of the evidence, a conclusion that the employer's proposed alternate methods, conditions, practices, operations, or processes would provide workers with protection that is at least equivalent to the protection afforded to them by the standard from which the employer is seeking the variance. (National defense variances do not require such a statement, and the statement submitted by an employer applying for a temporary variance must demonstrate that the employer is taking all available steps to safeguard workers.)

- Provide the employer's address, as well as the site location(s) that the variance will cover.

- A certification that the employer notified employees using the methods specified in the appropriate variance regulation.

- An original copy of the completed variance application signed by the employer or an authorized representative of the employer.

Submit the original of the completed application, as well as other relevant documents[1], to:

By regular mail:

Assistant Secretary for Occupational Safety and Health
Director Office of Technical Programs and Coordination Activities
Occupational Safety and Health Administration
U.S. Department of Labor
Room N3655
200 Constitution Avenue, NW
Washington, DC 20210

By facsimile:

202 693-1644

Electronic (email):

VarianceProgram@dol.gov

Experience in processing variance applications indicates that such applications are not appropriate in the following situations:

- The variance is from a "performance" standard, i.e., a standard that does not describe a specific method for meeting the requirements of the standard.

- The variance is from a "definition" in a standard, i.e., a provision that defines a term used in the standard, but does not expressly specify an action for meeting a requirement of the standard.

- The variance is a request for review and approval of a design or product developed for manufacture and commercial use.

- There is an OSHA standard in effect that allows the requested alternative.

- There exists an OSHA interpretation that permits the requested alternative.

- There is an updated edition of a national consensus or industry standard referenced in the OSHA standard, and that is the subject of the variance application, that permits the requested alternative.

- The application requests an exemption or exception from the requirements of the standard.

- If the application is for a temporary variance, the employer applied on or after the date the standard became effective.

- The applicant is contesting a citation involving the standard in question, or has an unresolved citation relating to this standard.

- The application involves locations that are solely within states or territories with OSHA-approved plans.

- The application is from a Federal agency.

Variance Application-Related Information

All information or documents submitted in an application becomes public unless the employer claims that some of the material consists of trade secrets[2] or confidential business information.[3] Employers seeking protection of trade secrets or other confidential business information in their variance application and supporting documents must include a request for such protection, as well as a justification for this request, in their variance application. Employers requesting such consideration should note that OSHA will assess the variance request solely on the basis of information that is available to the public.

Variance Application Checklists

To increase transparency and accessibility of variance related information, OSHA developed Variance Application Checklists designed to assist variance applicants to determine if their application for a variance is complete and appropriate.

- Permanent Variance Application Checklist [27 KB PDF*, 2 pages]

- Temporary Variance Application Checklist [27 KB PDF*, 2 pages]

- National Defense Variance Application Checklist [53 KB PDF*, 1 page]

- Experimental Variance Application Checklist [30 KB PDF*, 3 pages]

Variance Application Forms

The Variance Application forms are designed to assist prospective variance applicants to understand what information is required for a variance to be granted in a straight-forward, effective and user-friendly manner. Use of the Variance Application forms coupled with the Variance Application Checklists significantly reduces the burden of wading through the complexity of Federal Standards in order to interpret and understand the information obligations associated with applying for a variance.

- Permanent Variance Application Form [58 KB PDF*, 6 pages]

- Temporary Variance Application Form [62 KB PDF*, 7 pages]

- National Defense Variance Application Form [53 KB PDF*, 7 pages]

- Experimental Variance Application Form [60 KB PDF*, 6 pages]

Site Assessments

Either staff from OTPCA or the OSHA area office staff will perform an assessment of the employer's worksite when deemed necessary. OSHA will conduct site assessments when making decisions regarding the adequacy of an application and it needs further information to process an application. Site assessments are especially useful for temporary or experimental variances, or when OSHA receives employee complaints regarding a variance application. Generally, experimental variance applications will necessitate an onsite assessment to verify that the proposed experimental conditions are safe and healthful for workers. For temporary variances, the site assessment would investigate the availability of appropriate practices, means, methods, operations, and processes needed to come into compliance with the standard, as well as the ability of the employer to meet specific deadlines.

OSHA will arrange the site assessment with the employer in advance of its arrival. There are three parts to the subsequent site assessment: the opening conference, the site investigation, and the closing conference. Site assessments are not compliance inspections, and the OSHA compliance safety and health officers (CSHO) participating in a site assessment will not issue citations to the employer. However, the CSHO will inform the employer of any imminent dangers observed, and will request the employer to abate the hazard; if the employer refuses to do so, the CSHO will inform the nearest OSHA area office of the danger, and the area office will issue a citation. In addition, the site-assessment team will inform the employer of other hazards observed, and the need to abate the hazards, but will not issue citations for these hazards nor inform the area office of the hazards.

Granted or Denied Variances

After an employer submits a variance application, OSHA can either grant or deny the variance. Prior to granting a variance or interim order, OTPCA coordinates a thorough administrative and technical review of the application with other directorates and offices in OSHA, as well as its regional and area offices when appropriate. For permanent and experimental variances, the technical review determines if the alternate method proposed by the employer affords workers protection that is as effective as the protection that would result from complying with the standard from which the employer is seeking a variance. OSHA will deny a variance if the variance application fails the administrative or technical review process, including a failure to demonstrate that the proposed alternative would protect the employer's workers at least as effectively as the standard from which the employer is seeking the variance.

Copies of granted and denied variances, as well as interim orders, are available by accessing links from this Variance Web page.

Appendix G: Building an Inspection Checklist

Safety and health inspection checklists can take many forms; however, the primary purpose is to identify potential hazards in your workplace. Inspection documentation can take many forms, including the checklist identified below, a narrative format with explanation of each identified potential hazard, a graded checklist, a dated checklist wherein the number of times the potential hazard has been identified in previous inspections is identified, and whatever format is appropriate to your operations. It is important to conduct the site inspection and completion of the inspection documents on a frequent and consistent basis. The inspection document should be customized to the operation and ensure that the inspection encompasses *all* of the applicable standards as well as other potential hazards in the workplace.

SELF-INSPECTION

The most widely accepted way to identify hazards is to conduct safety and health inspections because the only way to be certain of an actual situation is to look at it directly from time to time.

Begin a program of self-inspection in your own workplace. Self-inspection is essential if you are to know where probable hazards exist and whether they are under control.

This section includes checklists designed to assist you in self-inspection fact-finding. The checklists can give you some indication of where to begin taking action to make your business safer and more healthful for all of your employees. These checklists are by no means all-inclusive, and not all of the checklists will apply to your business. You might want to start by selecting the areas that are most critical to your business, then expanding your self-inspection checklists over time to fully cover all areas that pertain to your business. Remember that a checklist is a tool to help, not a definitive statement of what is mandatory. Use checklists only for guidance.

Don't spend time with items that have no application to your business. Make sure that each item is seen by you or your designee, and leave nothing to memory or chance. Write down what you see or don't see and what you think you should do about it.

Add information from your completed checklists to injury information, employee information, and process and equipment information to build a foundation to help you determine what problems exist. Then, as you use the Occupational Safety and Health Administration (OSHA) standards in your problem-solving process, it will be easier for you to determine the actions needed to solve these problems.

Once the hazards have been identified, institute the control procedures described on page 9 and establish your 4-point safety and health program.

SELF-INSPECTION SCOPE

Your self-inspections should cover safety and health issues in the following areas:

- Processing, receiving, shipping, and storage: Equipment, job planning, layout, heights, floor loads, projection of materials, material handling and storage methods, training for material handling equipment.
- Building and grounds conditions: Floors, walls, ceilings, exits, stairs, walkways, ramps, platforms, driveways, aisles.
- Housekeeping program: Waste disposal, tools, objects, materials, leakage and spillage, cleaning methods, schedules, work areas, remote areas, storage areas.
- Electricity: Equipment, switches, breakers, fuses, switch boxes, junctions, special fixtures, circuits, insulation, extensions, tools, motors, grounding, national electric code compliance.
- Lighting: Type, intensity, controls, conditions, diffusion, location, glare, and shadow control.
- Heating and ventilation: Type, effectiveness, temperature, humidity, controls, natural and artificial ventilation, and exhausting.
- Machinery: Points of operation, flywheels, gears, shafts, pulleys, key ways, belts, couplings, sprockets, chains, frames, controls, lighting for tools and equipment, brakes, exhausting, feeding, oiling, adjusting, maintenance, lockout/tagout, grounding, work space, location, purchasing standards.
- Personnel: Training, including hazard identification training; experience; methods of checking machines before use; type of clothing; personal protective equipment (PPE); use of guards; tool storage; work practices; methods for cleaning, oiling, or adjusting machinery.
- Hand and power tools: Purchasing standards, inspection, storage, repair, types, maintenance, grounding, use and handling.
- Chemicals: Storage, handling, transportation, spills, disposals, amounts used, labeling, toxicity or other harmful effects, warning signs, supervision, training, protective clothing and equipment, hazard communication requirements.
- Fire prevention: Extinguishers, alarms, sprinklers, smoking rules, exits, personnel assigned, separation of flammable materials and dangerous operations, explosion-proof fixtures in hazardous locations, waste disposal, training of personnel.
- Maintenance: Provide regular and preventive maintenance on all equipment used at the worksite, recording all work performed on the machinery and by training personnel on the proper care and servicing of the equipment.
- PPE: Type, size, maintenance, repair, age, storage, assignment of responsibility, purchasing methods, standards observed, training in care and use, rules of use, method of assignment.
- Transportation: Motor vehicle safety, seat belts, vehicle maintenance, safe driver programs.

- First aid program/supplies: Medical care facilities locations, posted emergency phone numbers, accessible first aid kits.
- Evacuation plan: Establish and practice procedures for an emergency evacuation, e.g., fire, chemical/biological incidents, bomb threat; include escape procedures and routes, critical plant operations, employee accounting following an evacuation, rescue and medical duties, and ways to report emergencies.

SELF-INSPECTION CHECKLISTS

These checklists are by no means all-inclusive. You should add to them or delete items that do not apply to your business; however, carefully consider each item and then make your decision. You should refer to OSHA standards for specific guidance that may apply to your work situation. (Note: These checklists are typical for general industry but not for construction or maritime industries.)

Employer Posting

☐ Is the required OSHA job safety and health protection poster displayed in a prominent location where all employees are likely to see it?

☐ Are emergency telephone numbers posted where they can be readily found in case of emergency?

☐ Where employees may be exposed to toxic substances or harmful physical agents, has appropriate information concerning employee access to medical and exposure records and material safety data sheets (MSDSs) been posted or otherwise made readily available to affected employees?

☐ Are signs concerning exit routes, room capacities, floor loading, biohazards, exposures to x-ray, microwave, or other harmful radiation or substances posted where appropriate?

☐ Is the Summary of Work-Related Injuries and Illnesses (OSHA Form 300A) posted during the months of February, March, and April?

Recordkeeping

☐ Are occupational injuries or illnesses, except minor injuries requiring only first aid, recorded as required on the OSHA 300 log?

☐ Are employee medical records and records of employee exposure to hazardous substances or harmful physical agents up-to-date and in compliance with current OSHA standards?

☐ Are employee training records kept and accessible for review by employees, as required by OSHA standards?

☐ Have arrangements been made to retain records for the time period required for each specific type of record? (Some records must be maintained for at least 40 years.)

☐ Are operating permits and records up-to-date for items such as elevators, air pressure tanks, liquefied petroleum gas tanks, etc.?

OSHA Handbook for Small Businesses

☐ Do you have a working procedure to handle in-house employee complaints regarding safety and health?

☐ Are your employees advised of efforts and accomplishments of the safety and health program made to ensure they will have a workplace that is safe and healthful?

☐ Have you considered incentives for employees or workgroups who excel in reducing workplace injury and illnesses?

Medical Services and First Aid

☐ Is there a hospital, clinic, or infirmary for medical care near your workplace, or is at least one employee on each shift currently qualified to render first aid?

☐ Have all employees who are expected to respond to medical emergencies as part of their job responsibilities received first aid training, had the hepatitis B vaccination made available to them, had appropriate training on procedures to protect them from bloodborne pathogens, including universal precautions, and have available and understand how to use appropriate PPE to protect against exposure to bloodborne diseases?[a]

[a] Pursuant to an OSHA memorandum of July 1, 1992, employees who render first aid only as a collateral duty do not have to be offered a preexposure hepatitis B vaccine only if the employer includes and implements the following requirements in its exposure control plan: (1) the employer must record all first aid incidents involving the presence of blood or other potentially infectious materials before the end of the work shift during which the first aid incident occurred; (2) the employer must comply with postexposure evaluation, prophylaxis, and follow-up requirements of the bloodborne pathogens standard with respect to exposure incidents, as defined by the standard; (3) the employer must train designated first aid providers about the reporting procedure; and (4) the employer must offer to initiate the hepatitis B vaccination series within 24 hours to all unvaccinated first aid providers who have rendered assistance in any situation involving the presence of blood or other potentially infectious materials.

☐ If employees have had an exposure incident involving bloodborne pathogens, was an immediate postexposure medical evaluation and follow-up provided?

☐ Are medical personnel readily available for advice and consultation on matters of employees' health?

☐ Are emergency phone numbers posted?

☐ Are fully supplied first aid kits easily accessible to each work area, periodically inspected, and replenished as needed?

☐ Have first aid kits and supplies been approved by a physician, indicating that they are adequate for a particular area or operation?

☐ Is there an eyewash station or sink available for quick drenching or flushing of the eyes and body in areas where corrosive liquids or materials are handled?

Fire Protection

☐ Is your local fire department familiar with your facility, its location, and specific hazards?

☐ If you have a fire alarm system, is it certified as required and tested annually?

☐ If you have interior standpipes and valves, are they inspected regularly?

☐ If you have exterior standpipes and valves, are they inspected regularly?

☐ If you have outside private fire hydrants, are they flushed at least once a year and on a routine preventive maintenance schedule?

☐ Are fire doors and shutters in good operating condition?

☐ Are fire doors and shutters unobstructed and protected against obstructions, including their counterweights?

☐ Are fire door and shutter fusible links in place?

☐ Are automatic sprinkler system water control valves, air, and water pressure checked periodically as required?

☐ Is the maintenance of automatic sprinkler systems assigned to responsible persons or to a sprinkler contractor?

☐ Are sprinkler heads protected by metal guards if exposed to potential physical damage?

☐ Is proper clearance maintained below sprinkler heads?

☐ Are portable fire extinguishers provided in adequate number and type and mounted in readily accessible locations?

☐ Are fire extinguishers recharged regularly with this noted on the inspection tag?

☐ Are employees periodically instructed in the use of fire extinguishers and fire protection procedures?

Personal Protective Equipment and Clothing

☐ Has the employer determined whether hazards that require the use of PPE (e.g., head, eye, face, hand, or foot protection) are present or are likely to be present?

☐ If hazards or the likelihood of hazards is found, are employers selecting appropriate and properly fitted PPE suitable for protection from these hazards and ensuring that affected employees use it?

☐ Have both the employer and the employees been trained on PPE procedures, i.e., what PPE is necessary for job tasks, when workers need it, and how to properly wear and adjust it?

☐ Are protective goggles or face shields provided and worn where there is any danger of flying particles or corrosive materials?

☐ Are approved safety glasses required to be worn at all times in areas where there is a risk of eye injuries such as punctures, abrasions, contusions, or burns?

☐ Are employees who wear corrective lenses (glasses or contacts) in workplaces with harmful exposures required to wear only approved safety glasses, protective goggles, or use other medically approved precautionary procedures?

☐ Are protective gloves, aprons, shields, or other means provided and required where employees could be cut or where there is reasonably anticipated exposure to corrosive liquids, chemicals, blood, or other potentially infectious materials? See the OSHA bloodborne pathogens standard, 29 CFR 1910.1030(b), for the definition of "other potentially infectious materials."

☐ Are hard hats required, provided, and worn where danger of falling objects exists?

☐ Are hard hats periodically inspected for damage to the shell and suspension system?

☐ Is appropriate foot protection required where there is the risk of foot injuries from hot, corrosive, or poisonous substances, falling objects, crushing, or penetrating actions?

☐ Are approved respirators provided when needed? (See 29 CFR 1910.134 for detailed information on respirators or check OSHA's website.)

☐ Is all PPE maintained in a sanitary condition and ready for use?

☐ Are food or beverages consumed only in areas where there is no exposure to toxic material, blood, or other potentially infectious materials?

☐ Is protection against the effects of occupational noise provided when sound levels exceed those of the OSHA noise standard?

☐ Are adequate work procedures, PPE, and other equipment provided and used when cleaning up spilled hazardous materials?

☐ Are appropriate procedures in place to dispose of or decontaminate PPE contaminated with, or reasonably anticipated to be contaminated with, blood or other potentially infectious materials?

General Work Environment

☐ Are all worksites clean, sanitary, and orderly?

☐ Are work surfaces kept dry and appropriate means taken to assure the surfaces are slip resistant?

☐ Are all spilled hazardous materials or liquids, including blood and other potentially infectious materials, cleaned up immediately and according to proper procedures?

☐ Is combustible scrap, debris, and waste stored safely and removed from the worksite promptly?

☐ Is all regulated waste, as defined in the OSHA bloodborne pathogens standard (29 CFR 1910.1030), discarded according to federal, state, and local regulations?

☐ Are accumulations of combustible dust routinely removed from elevated surfaces, including the overhead structure of buildings, etc.?

☐ Is combustible dust cleaned up with a vacuum system to prevent suspension of dust particles in the environment?

☐ Is metallic or conductive dust prevented from entering or accumulating on or around electrical enclosures or equipment?

☐ Are covered metal waste cans used for oily or paint-soaked waste?

☐ Are all oil and gas-fired devices equipped with flame failure controls to prevent flow of fuel if pilots or main burners are not working?

☐ Are paint spray booths, dip tanks, etc., cleaned regularly?

☐ Are the minimum number of toilets and washing facilities provided and maintained in a clean and sanitary fashion?

☐ Are all work areas adequately illuminated?

☐ Are pits and floor openings covered or otherwise guarded?

☐ Have all confined spaces been evaluated for compliance with 29 CFR 1910.146 (permit-required confined spaces)?

Walkways

☐ Are aisles and passageways kept clear and marked as appropriate?

☐ Are wet surfaces covered with nonslip materials?

☐ Are holes in the floor, sidewalk, or other walking surface repaired properly, covered, or otherwise made safe?

☐ Is there safe clearance for walking in aisles where motorized or mechanical handling equipment is operating?

☐ Are materials or equipment stored in such a way that sharp projections will not interfere with the walkway?

☐ Are spilled materials cleaned up immediately?

☐ Are changes of direction or elevations readily identifiable?

☐ Are aisles or walkways that pass near moving or operating machinery, welding operations, or similar operations arranged so employees will not be subjected to potential hazards?

☐ Is adequate headroom provided for the entire length of any aisle or walkway?

☐ Are standard guardrails provided wherever aisle or walkway surfaces are elevated more than 30 inches (76.20 centimeters) above any adjacent floor or the ground?

☐ Are bridges provided over conveyors and similar hazards?

Floor and Wall Openings

☐ Are floor openings guarded by a cover, a guardrail, or equivalent on all sides (except at stairways or ladder entrances)?

☐ Are toeboards installed around the edges of permanent floor openings where persons may pass below the opening?

☐ Are skylight screens able to withstand a load of at least 200 pounds (90.7 kilograms)?

☐ Is the glass in windows, doors, glass walls, etc., subject to possible human impact, of sufficient thickness and type for the condition of use?

☐ Are grates or similar type covers over floor openings such as floor drains designed to allow unimpeded foot traffic or rolling equipment?

☐ Are unused portions of service pits and pits not in use either covered or protected by guardrails or equivalent?

☐ Are manhole covers, trench covers and similar covers, and their supports designed to carry a truck rear axle load of at least 20,000 pounds (9072 kilograms) when located in roadways and subject to vehicle traffic?

☐ Are floor or wall openings in fire-resistant construction provided with doors or covers compatible with the fire rating of the structure and provided with a self-closing feature when appropriate?

Stairs and Stairways

☐ Do standard stair rails or handrails on all stairways have at least four risers?

☐ Are all stairways at least 22 inches (55.88 centimeters) wide?

☐ Do stairs have landing platforms not less than 30 inches (76.20 centimeters) in the direction of travel and extend 22 inches (55.88 centimeters) in width at every 12 feet (3.6576 meters) or less of vertical rise?

☐ Do stairs angle no more than 50 and no less than 30 degrees?

☐ Are stairs of hollow-pan-type treads and landings filled to the top edge of the pan with solid material?

☐ Are step risers on stairs uniform from top to bottom?

☐ Are steps slip resistant?

☐ Are stairway handrails located between 30 inches (76.20 centimeters) and 34 inches (86.36 centimeters) above the leading edge of stair treads?

☐ Do stairway handrails have at least 3 inches (7.62 centimeters) of clearance between the handrails and the wall or surface they are mounted on?

☐ Where doors or gates open directly on a stairway, is a platform provided so the swing of the door does not reduce the width of the platform to less than 21 inches (53.34 centimeters)?

☐ Are stairway handrails capable of withstanding a load of 200 pounds (90.7 kilograms), applied within 2 inches (5.08 centimeters) of the top edge in any downward or outward direction?

☐ Where stairs or stairways exit directly into any area where vehicles may be operated, are adequate barriers and warnings provided to prevent employees from stepping into the path of traffic?

☐ Do stairway landings have a dimension measured in the direction of travel at least equal to the width of the stairway?

☐ Is the vertical distance between stairway landings limited to 12 feet (3.6576 meters) or less?

Elevated Surfaces

☐ Are signs posted, when appropriate, showing the elevated surface load capacity?

☐ Are surfaces that are elevated more than 30 inches (76.20 centimeters) provided with standard guardrails?

☐ Are surfaces that are elevated more than 30 inches (76.20 centimeters) provided with standard guardrails?

☐ Are all elevated surfaces beneath which people or machinery could be exposed to falling objects provided with standard 4-inch (10.16-centimeter) toeboards?

☐ Is a permanent means of access and egress provided to elevated storage and work surfaces?

☐ Is required headroom provided where necessary?

☐ Is material on elevated surfaces piled, stacked, or racked in a manner to prevent it from tipping, falling, collapsing, rolling, or spreading?

☐ Are dock boards or bridge plates used when transferring materials between docks and trucks or railcars?

Exiting or Egress—Evacuation

☐ Are all exits marked with an exit sign and illuminated by a reliable light source?

☐ Are the directions to exits, when not immediately apparent, marked with visible signs?

☐ Are doors, passageways, or stairways that are neither exits nor access to exits, but could be mistaken for exits, appropriately marked "NOT AN EXIT," "TO BASEMENT," "STOREROOM," etc.?

- [ ] Are exit signs labeled with the word *EXIT* in lettering at least 5 inches (12.70 centimeters) high and the stroke of the lettering at least 1/2 inch (1.2700 centimeters) wide?
- [ ] Are exit doors side hinged?
- [ ] Are all exits kept free of obstructions?
- [ ] Are at least two means of egress provided from elevated platforms, pits, or rooms where the absence of a second exit would increase the risk of injury from hot, poisonous, corrosive, suffocating, flammable, or explosive substances?
- [ ] Are there sufficient exits to permit prompt escape in case of emergency?
- [ ] Are special precautions taken to protect employees during construction and repair operations?
- [ ] Are the number of exits from each floor of a building and the number of exits from the building itself appropriate for the building occupancy load?
- [ ] Are exit stairways that are required to be separated from other parts of a building enclosed by at least 2-hour fire-resistive construction in buildings more than four stories in height, and not less than 1-hour fire-resistive construction elsewhere?
- [ ] Where ramps are used as part of required exiting from a building, is the ramp slope limited to 1 foot (0.3048 meter) vertical and 12 feet (3.6576 meters) horizontal?
- [ ] Where exiting will be through frameless glass doors, glass exit doors, storm doors, etc., are the doors fully tempered and meet the safety requirements for human impact?

Exit Doors

- [ ] Are doors that are required to serve as exits designed and constructed so that the path of exit travel is obvious and direct?
- [ ] Are windows that could be mistaken for exit doors made inaccessible by means of barriers or railings?
- [ ] Are exit doors able to be opened from the direction of exit travel without the use of a key or any special knowledge or effort when the building is occupied?
- [ ] Is a revolving, sliding, or overhead door prohibited from serving as a required exit door?
- [ ] Where panic hardware is installed on a required exit door, will it allow the door to open by applying a force of 15 pounds (6.80 kilograms) or less in the direction of the exit traffic?
- [ ] Are doors on cold storage rooms provided with an inside release mechanism that will release the latch and open the door even if the door is padlocked or otherwise locked on the outside?
- [ ] Where exit doors open directly onto any street, alley, or other area where vehicles may be operated, are adequate barriers and warnings provided to prevent employees from stepping into the path of traffic?
- [ ] Are doors that swing in both directions and are located between rooms where there is frequent traffic provided with viewing panels in each door?

Portable Ladders

- [ ] Are all ladders maintained in good condition, joints between steps and side rails tight, all hardware and fittings securely attached, and moveable parts operating freely without binding or undue play?
- [ ] Are nonslip safety feet provided on each metal or rung ladder, and are ladder rungs and steps free of grease and oil?
- [ ] Are employees prohibited from placing a ladder in front of doors opening toward the ladder unless the door is blocked open, locked, or guarded?
- [ ] Are employees prohibited from placing ladders on boxes, barrels, or other unstable bases to obtain additional height?
- [ ] Are employees required to face the ladder when ascending or descending?
- [ ] Are employees prohibited from using ladders that are broken, have missing steps, rungs, or cleats, broken side rails, or other faulty equipment?

☐ Are employees instructed not to use the top step of ordinary stepladders as a step?

☐ When portable rung ladders are used to gain access to elevated platforms, roofs, etc., does the ladder always extend at least 3 feet (0.9144 meters) above the elevated surface?

☐ Are employees required to secure the base of a portable rung or cleat-type ladder to prevent slipping, or otherwise lash or hold it in place?

☐ Are portable metal ladders legibly marked with signs reading "CAUTION—Do Not Use Around Electrical Equipment" or equivalent wording?

☐ Are employees prohibited from using ladders as guys, braces, skids, gin poles, or for other than their intended purposes?

☐ Are employees instructed to only adjust extension ladders while standing at a base (not while standing on the ladder or from a position above the ladder)?

☐ Are metal ladders inspected for damage?

☐ Are the rungs of ladders uniformly spaced at 12 inches (30.48 centimeters) center to center?

Hand Tools and Equipment

☐ Are all tools and equipment (both company and employee owned) used at the workplace in good condition?

☐ Are hand tools, such as chisels, punches, etc., which develop mushroomed heads during use, reconditioned or replaced as necessary?

☐ Are broken or fractured handles on hammers, axes, and similar equipment replaced promptly?

☐ Are worn or bent wrenches replaced?

☐ Are appropriate handles used on files and similar tools?

☐ Are employees aware of hazards caused by faulty or improperly used hand tools?

☐ Are appropriate safety glasses, face shields, etc., used while using hand tools or equipment that might produce flying materials or be subject to breakage?

☐ Are jacks checked periodically to ensure they are in good operating condition?

☐ Are tool handles wedged tightly into the heads of all tools?

☐ Are tool cutting edges kept sharp so the tool will move smoothly without binding or skipping?

☐ Are tools stored in a dry, secure location where they cannot be tampered with?

☐ Is eye and face protection used when driving hardened or tempered studs or nails?

Portable (Power-Operated) Tools and Equipment

☐ Are grinders, saws, and similar equipment provided with appropriate safety guards?

☐ Are power tools used with proper shields, guards, or attachments, as recommended by the manufacturer?

☐ Are portable circular saws equipped with guards above and below the base shoe?

☐ Are circular saw guards checked to ensure that they are not wedged up, leaving the lower portion of the blade unguarded?

☐ Are rotating or moving parts of equipment guarded to prevent physical contact?

☐ Are all cord-connected, electrically operated tools and equipment effectively grounded or of the approved double insulated type?

☐ Are effective guards in place over belts, pulleys, chains, and sprockets on equipment such as concrete mixers, air compressors, etc.?

☐ Are portable fans provided with full guards or screens having openings 1/2 inch (1.2700 centimeters) or less?

☐ Is hoisting equipment available and used for lifting heavy objects, and are hoist ratings and characteristics appropriate for the task?

☐ Are ground-fault circuit interrupters provided on all temporary electrical 15- and 20-ampere circuits used during periods of construction?

☐ Are pneumatic and hydraulic hoses on powder-operated tools checked regularly for deterioration or damage?

Abrasive Wheel Equipment Grinders

☐ Is the work rest used and kept adjusted to within 1/8 inch (0.3175 centimeter) of the wheel?

☐ Is the adjustable tongue on the top side of the grinder used and kept adjusted to within 1/4 inch (0.6350 centimeter) of the wheel?

☐ Do side guards cover the spindle, nut, and flange and 75 percent of the wheel diameter?

☐ Are bench and pedestal grinders permanently mounted?

☐ Are goggles or face shields always worn when grinding?

☐ Is the maximum revolutions per minute (rpm) rating of each abrasive wheel compatible with the rpm rating of the grinder motor?

☐ Are fixed or permanently mounted grinders connected to their electrical supply system with metallic conduit or another permanent wiring method?

☐ Does each grinder have an individual on and off control switch?

☐ Is each electrically operated grinder effectively grounded?

☐ Are new abrasive wheels visually inspected and ring tested before they are mounted?

☐ Are dust collectors and powered exhausts provided on grinders used in operations that produce large amounts of dust?

☐ Are splash guards mounted on grinders that use coolant to prevent the coolant from reaching employees?

☐ Is cleanliness maintained around grinders?

Powder-Actuated Tools

☐ Are employees who operate powder-actuated tools trained in their use and required to carry a valid operator's card?

☐ Is each powder-actuated tool stored in its own locked container when not being used?

☐ Is a sign at least 7 inches (17.78 centimeters) by 10 inches (25.40 centimeters) with bold face type reading "POWDER-ACTUATED TOOL IN USE" conspicuously posted when the tool is being used?

☐ Are powder-actuated tools left unloaded until they are ready to be used?

☐ Are powder-actuated tools inspected for obstructions or defects each day before use?

☐ Do powder-actuated tool operators have and use appropriate PPE, such as hard hats, safety goggles, safety shoes, and ear protectors?

Machine Guarding

☐ Is there a training program to instruct employees on safe methods of machine operation?

☐ Is there adequate supervision to ensure that employees are following safe machine operating procedures?

☐ Is there a regular program of safety inspection of machinery and equipment?

☐ Are all machinery and equipment kept clean and properly maintained?

☐ Is sufficient clearance provided around and between machines to allow for safe operations, setup and servicing, material handling, and waste removal?

☐ Are equipment and machinery securely placed and anchored to prevent tipping or other movement that could result in personal injury?

☐ Is there a power shutoff switch within reach of the operator's position at each machine?

☐ Can electric power to each machine be locked out for maintenance, repair, or security?

☐ Are the non-current-carrying metal parts of electrically operated machines bonded and grounded?

☐ Are foot-operated switches guarded or arranged to prevent accidental actuation by personnel or falling objects?

☐ Are manually operated valves and switches controlling the operation of equipment and machines clearly identified and readily accessible?

☐ Are all emergency stop buttons colored red?

☐ Are all pulleys and belts within 7 feet (2.1336 meters) of the floor or working level properly guarded?

☐ Are all moving chains and gears properly guarded?

☐ Are splash guards mounted on machines that use coolant to prevent the coolant from reaching employees?

☐ Are methods provided to protect the operator and other employees in the machine area from hazards created at the point of operation, ingoing nip points, rotating parts, flying chips, and sparks?

☐ Are machine guards secure and arranged so they do not cause a hazard while in use?

☐ If special hand tools are used for placing and removing material, do they protect the operator's hands?

☐ Are revolving drums, barrels, and containers guarded by an enclosure that is interlocked with the drive mechanism so that revolution cannot occur unless the guard enclosure is in place?

☐ Do arbors and mandrels have firm and secure bearings, and are they free from play?

☐ Are provisions made to prevent machines from automatically starting when power is restored after a power failure or shutdown?

☐ Are machines constructed so as to be free from excessive vibration when the largest size tool is mounted and run at full speed?

☐ If machinery is cleaned with compressed air, is air pressure controlled and PPE or other safeguards utilized to protect operators and other workers from eye and body injury?

☐ Are fan blades protected with a guard having openings no larger than 1/2 inch (1.2700 centimeters) when operating within 7 feet (2.1336 meters) of the floor?

☐ Are saws used for ripping equipped with antikickback devices and spreaders?

☐ Are radial arm saws so arranged that the cutting head will gently return to the back of the table when released?

Lockout/Tagout Procedures

☐ Is all machinery or equipment capable of movement required to be deenergized or disengaged and blocked or locked out during cleaning, servicing, adjusting, or setting up operations?

☐ If the power disconnect for equipment does not also disconnect the electrical control circuit, are the appropriate electrical enclosures identified, and is a means provided to ensure that the control circuit can also be disconnected and locked out?

☐ Is the locking out of control circuits instead of locking out main power disconnects prohibited?

☐ Are all equipment control valve handles provided with a means for locking out?

☐ Does the lockout procedure require that stored energy (mechanical, hydraulic, air, etc.) be released or blocked before equipment is locked out for repairs?

☐ Are appropriate employees provided with individually keyed personal safety locks?

☐ Are employees required to keep personal control of their key(s) while they have safety locks in use?

☐ Is it required that only the employee exposed to the hazard can place or remove the safety lock?

☐ Is it required that employees check the safety of the lockout by attempting a start-up after making sure no one is exposed?

☐ Are employees instructed to always push the control circuit stop button prior to reenergizing the main power switch?

☐ Is there a means provided to identify any or all employees who are working on locked-out equipment by their locks or accompanying tags?

☐ Are a sufficient number of accident prevention signs or tags and safety padlocks provided for any reasonably foreseeable repair emergency?

☐ When machine operations, configuration, or size require an operator to leave the control station and part of the machine could move if accidentally activated, is the part required to be separately locked out or blocked?

☐ If equipment or lines cannot be shut down, locked out, and tagged, is a safe job procedure established and rigidly followed?

Welding, Cutting, and Brazing

☐ Are only authorized and trained personnel permitted to use welding, cutting, or brazing equipment?

☐ Does each operator have a copy of and follow the appropriate operating instructions?

☐ Are compressed gas cylinders regularly examined for obvious signs of defects, deep rusting, or leakage?

☐ Is care used in handling and storage of cylinders, safety valves, relief valves, etc., to prevent damage?

☐ Are precautions taken to prevent the mixture of air or oxygen with flammable gases, except at a burner or in a standard torch?

☐ Are only approved apparatuses (torches, regulators, pressure-reducing valves, acetylene generators, manifolds) used?

☐ Are cylinders kept away from sources of heat and elevators, stairs, or gangways?

☐ Is it prohibited to use cylinders as rollers or supports?

☐ Are empty cylinders appropriately marked and their valves closed?

☐ Are signs posted reading "DANGER, NO SMOKING, MATCHES, OR OPEN LIGHTS" or the equivalent?

☐ Are cylinders, cylinder valves, couplings, regulators, hoses, and apparatuses kept free of oily or greasy substances?

☐ Is care taken not to drop or strike cylinders?

☐ Are regulators removed and valve protection caps put in place before moving cylinders, unless they are secured on special trucks?

☐ Do cylinders without fixed wheels have keys, handles, or nonadjustable wrenches on stem valves when in service?

☐ Are liquefied gases stored and shipped valve end up with valve covers in place?

☐ Are employees trained never to crack a fuel gas cylinder valve near sources of ignition?

☐ Before a regulator is removed, is the valve closed and gas released?

☐ Is red used to identify the acetylene (and other fuel gas) hose, green for the oxygen hose, and black for inert gas and air hoses?

☐ Are pressure-reducing regulators used only for the gases and pressures for which they are intended?

☐ Is open-circuit (no-load) voltage of arc welding and cutting machines as low as possible and not in excess of the recommended limits?

☐ Under wet conditions, are automatic controls for reducing no-load voltage used?

☐ Is grounding of the machine frame and safety ground connections of portable machines checked periodically?

☐ Are electrodes removed from the holders when not in use?

☐ Is it required that electric power to the welder be shut off when no one is in attendance?

☐ Is suitable fire extinguishing equipment available for immediate use?

☐ Is the welder forbidden to coil or loop welding electrode cable around his body?

☐ Are wet machines thoroughly dried and tested before use?

☐ Are work and electrode lead cables frequently inspected for wear and damage, and replaced when needed?

☐ Are cable connectors adequately insulated?

☐ When the object to be welded cannot be moved and fire hazards cannot be removed, are shields used to confine heat, sparks, and slag?

☐ Are fire watchers assigned when welding or cutting is performed in locations where a serious fire might develop?

☐ Are combustible floors kept wet, covered with damp sand, or protected by fire-resistant shields?

☐ Are personnel protected from possible electrical shock when floors are wet?

☐ Are precautions taken to protect combustibles on the other side of metal walls when welding is underway?

☐ Are used drums, barrels, tanks, and other containers thoroughly cleaned of substances that could explode, ignite, or produce toxic vapors before hot work begins?

☐ Do eye protection, helmets, hand shields, and goggles meet appropriate standards?

☐ Are employees exposed to the hazards created by welding, cutting, or brazing operations protected with PPE and clothing?

☐ Is a check made for adequate ventilation in and where welding or cutting is performed?

☐ When working in confined places, are environmental monitoring tests done and means provided for quick removal of welders in case of an emergency?

Compressors and Compressed Air

☐ Are compressors equipped with pressure relief valves and pressure gauges?

☐ Are compressor air intakes installed and equipped so as to ensure that only clean, uncontaminated air enters the compressor?

☐ Are air filters installed on the compressor intake?

☐ Are compressors operated and lubricated in accordance with the manufacturer's recommendations?

☐ Are safety devices on compressed air systems checked frequently?

☐ Before a compressor's pressure system is repaired, is the pressure bled off and the system locked out?

☐ Are signs posted to warn of the automatic starting feature of the compressors?

☐ Is the belt drive system totally enclosed to provide protection for the front, back, top, and sides?

☐ Are employees strictly prohibited from directing compressed air toward a person?

☐ Are employees prohibited from using highly compressed air for cleaning purposes?

☐ When compressed air is used to clean clothing, are employees trained to reduce the pressure to less than 10 pounds per square inch (psi)?

☐ When using compressed air for cleaning, do employees wear protective chip guarding and PPE?

☐ Are safety chains or other suitable locking devices used at couplings of high-pressure hose lines where a connection failure would create a hazard?

☐ Before compressed air is used to empty containers of liquid, is the safe working pressure of the container checked?

☐ When compressed air is used with abrasive blast cleaning equipment, is the operating valve a type that must be held open manually?

☐ When compressed air is used to inflate auto tires, are a clip-on chuck and an inline regulator preset to 40 psi required?

☐ Are employees prohibited from using compressed air to clean up or move combustible dust if such action could cause the dust to be suspended in the air and cause a fire or explosion hazard?

Compressors/Air Receivers

☐ Is every receiver equipped with a pressure gauge and one or more automatic, spring-loaded safety valves?

☐ Is the total relieving capacity of the safety valve able to prevent pressure in the receiver from exceeding the maximum allowable working pressure of the receiver by more than 10 percent?

☐ Is every air receiver provided with a drain pipe and valve at the lowest point for the removal of accumulated oil and water?

☐ Are compressed air receivers periodically drained of moisture and oil?

☐ Are all safety valves tested at regular intervals to determine whether they are in good operating condition?

☐ Is there a current operating permit?

☐ Is the inlet of air receivers and piping systems kept free of accumulated oil and carbonaceous materials?

Compressed Gas Cylinders

☐ Are cylinders with a water weight capacity over 30 pounds (13.6 kilograms) equipped with a means to connect a valve protector device, or with a collar or recess to protect the valve?

☐ Are cylinders legibly marked to clearly identify the type of gas?

☐ Are compressed gas cylinders stored in areas protected from external heat sources such as flame impingement, intense radiant heat, electric arcs, or high-temperature lines?

☐ Are cylinders located or stored in areas where they will not be damaged by passing or falling objects or subject to tampering by unauthorized persons?

☐ Are cylinders stored or transported in a manner to prevent them from creating a hazard by tipping, falling, or rolling?

☐ Are cylinders containing liquefied fuel gas stored or transported in a position so that the safety relief device is always in direct contact with the vapor space in the cylinder?

☐ Are valve protectors always placed on cylinders when the cylinders are not in use or connected for use?

☐ Are all valves closed off before a cylinder is moved, when the cylinder is empty and at the completion of each job?

☐ Are low-pressure fuel gas cylinders checked periodically for corrosion, general distortion, cracks, or any other defect that might indicate a weakness or render them unfit for service?

☐ Does the periodic check of low-pressure fuel gas cylinders include a close inspection of the cylinders' bottoms?

Hoist and Auxiliary Equipment

☐ Is each overhead electric hoist equipped with a limit device to stop the hook at its highest and lowest point of safe travel?

☐ Will each hoist automatically stop and hold any load up to 125 percent of its rated load if its actuating force is removed?

☐ Is the rated load of each hoist legibly marked and visible to the operator?

☐ Are stops provided at the safe limits of travel for trolley hoists?

☐ Are the controls of hoists plainly marked to indicate the direction of travel or motion?

☐ Is each cage-controlled hoist equipped with an effective warning device?

☐ Are close-fitting guards or other suitable devices installed on each hoist to ensure that hoist ropes will be maintained in the sheave grooves?

☐ Are all hoist chains or ropes long enough to handle the full range of movement of the application while maintaining two full wraps around the drum at all times?

☐ Are guards provided for nip points or contact points between hoist ropes and sheaves permanently located within 7 feet (2.1336 meters) of the floor, ground, or working platform?

☐ Are employees prohibited from using chains or rope slings that are kinked or twisted and prohibited from using the hoist rope or chain wrapped around the load as a substitute for a sling?

☐ Is the operator instructed to avoid carrying loads above people?

Industrial Trucks—Forklifts

☐ Are employees properly trained in the use of the type of industrial truck they operate?

☐ Are only trained personnel allowed to operate industrial trucks?

☐ Is substantial overhead protective equipment provided on high lift rider equipment?

☐ Are the required lift truck operating rules posted and enforced?

☐ Is directional lighting provided on each industrial truck that operates in an area with less than 2 foot-candles per square foot of general lighting?

☐ Does each industrial truck have a warning horn, whistle, gong, or other device that can be clearly heard above normal noise in the areas where it is operated?

☐ Are the brakes on each industrial truck capable of bringing the vehicle to a complete and safe stop when fully loaded?

☐ Does the parking brake of the industrial truck prevent the vehicle from moving when unattended?

☐ Are industrial trucks that operate where flammable gases, vapors, combustible dust, or ignitable fibers may be present approved for such locations?

☐ Are motorized hand and hand/rider trucks designed so that the brakes are applied and power to the drive motor shuts off when the operator releases his or her grip on the device that controls the truck's travel?

☐ Are industrial trucks with internal combustion engines that are operated in buildings or enclosed areas carefully checked to ensure that such operations do not cause harmful concentrations of dangerous gases or fumes?

☐ Are safe distances maintained from the edges of elevated ramps and platforms?

☐ Are employees prohibited from standing or passing under elevated portions of trucks, whether loaded or empty?

☐ Are unauthorized employees prohibited from riding on trucks?

☐ Are operators prohibited from driving up to anyone standing in front of a fixed object?

☐ Are arms and legs kept inside the running lines of the truck?

☐ Are loads handled only within the rated capacity of the truck?

☐ Are trucks in need of repair removed from service immediately?

Spraying Operations

☐ Is adequate ventilation provided before spraying operations are started?

☐ Is mechanical ventilation provided when spraying operations are performed in enclosed areas?

☐ When mechanical ventilation is provided during spraying operations, is it so arranged that it will not circulate the contaminated air?

☐ Is the spray area free of hot surfaces and at least 20 feet (6.096 meters) from flames, sparks, operating electrical motors, and other ignition sources?

☐ Are portable lamps used to illuminate spray areas suitable for use in a hazardous location?

☐ Is approved respiratory equipment provided and used when appropriate during spraying operations?

☐ Do solvents used for cleaning have a flash point to 100 degrees Fahrenheit or more?

☐ Are fire control sprinkler heads kept clean?

☐ Are "NO SMOKING" signs posted in spray areas, paint rooms, paint booths, and paint storage areas?

☐ Is the spray area kept clean of combustible residue?

☐ Are spray booths constructed of metal, masonry, or other substantial noncombustible material?

☐ Are spray booth floors and baffles noncombustible and easily cleaned?

☐ Is infrared drying apparatus kept out of the spray area during spraying operations, and is the spray booth completely ventilated before using the drying apparatus?

☐ Is the electric drying apparatus properly grounded?

☐ Are lighting fixtures for spray booths located outside the booth with the interior lighted through sealed clear panels?

☐ Are the electric motors for exhaust fans placed outside booths or ducts?

☐ Are belts and pulleys inside the booth fully enclosed?

☐ Do ducts have access doors to allow cleaning?

☐ Do all drying spaces have adequate ventilation?

Entering Confined Spaces

☐ Are confined spaces thoroughly emptied of any corrosive or hazardous substances, such as acids or caustics, before entry?

☐ Are all lines to a confined space that contain inert, toxic, flammable, or corrosive materials valved off and blanked or disconnected and separated before entry?

☐ Are all impellers, agitators, or other moving parts and equipment inside confined spaces locked out if they present a hazard?

☐ Is either natural or mechanical ventilation provided prior to confined space entry?

☐ Are appropriate atmospheric tests performed to check for oxygen deficiency, toxic substances, and explosive concentrations in the confined space before entry?

☐ Is adequate illumination provided for the work to be performed in the confined space?

☐ Is the atmosphere inside the confined space frequently tested or continuously monitored during work?

☐ Is there a trained and equipped standby employee positioned outside the confined space, whose sole responsibility is to watch the work in progress, sound an alarm if necessary, and render assistance?

☐ Is the standby employee appropriately trained and equipped to handle an emergency?

☐ Are employees prohibited from entering the confined space without lifelines and respiratory equipment if there is any question as to the cause of an emergency?

☐ Is approved respiratory equipment required if the atmosphere inside the confined space cannot be made acceptable?

☐ Is all portable electrical equipment used inside confined spaces either grounded and insulated or equipped with ground fault protection?

☐ Are compressed gas bottles forbidden inside the confined space?

☐ Before gas welding or burning is started in a confined space, are hoses checked for leaks, torches lighted only outside the confined area, and the confined area tested for an explosive atmosphere each time before a lighted torch is taken into the confined space?

☐ If employees will be using oxygen-consuming equipment such as salamanders, torches, furnaces, etc., in a confined space, is sufficient air provided to assure combustion without reducing the oxygen concentration of the atmosphere below 19.5 percent by volume?

☐ Whenever combustion-type equipment is used in a confined space, are provisions made to ensure the exhaust gases are vented outside of the enclosure?

☐ Is each confined space checked for decaying vegetation or animal matter that may produce methane?

☐ Is the confined space checked for possible industrial waste that could contain toxic properties?

☐ If the confined space is below ground and near areas where motor vehicles will be operating, is it possible for vehicle exhaust or carbon monoxide to enter the space?

Environmental Controls

☐ Are all work areas properly illuminated?

☐ Are employees instructed in proper first aid and other emergency procedures?

☐ Are hazardous substances, blood, and other potentially infectious materials that may cause harm by inhalation, ingestion, or skin absorption or contact identified?

☐ Are employees aware of the hazards involved with the various chemicals they may be exposed to in their work environment, such as ammonia, chlorine, epoxies, caustics, etc.?

☐ Is employee exposure to chemicals in the workplace kept within acceptable levels?

☐ Can a less harmful method or product be used?

☐ Is the work area ventilation system appropriate for the work performed?

☐ Are spray painting operations performed in spray rooms or booths equipped with an appropriate exhaust system?

☐ Is employee exposure to welding fumes controlled by ventilation, use of respirators, exposure time limits, or other means?

☐ Are welders and other nearby workers provided with flash shields during welding operations?

☐ If forklifts and other vehicles are used in buildings or other enclosed areas, are the carbon monoxide levels kept below maximum acceptable concentration?

☐ Has there been a determination that noise levels in the facilities are within acceptable levels?

☐ Are steps being taken to use engineering controls to reduce excessive noise levels?

☐ Are proper precautions being taken when handling asbestos and other fibrous materials?

☐ Are caution labels and signs used to warn of hazardous substances (e.g., asbestos) and biohazards (e.g., bloodborne pathogens)?

☐ Are wet methods used, when practicable, to prevent the emission of airborne asbestos fibers, silica dust, and similar hazardous materials?

☐ Are engineering controls examined and maintained or replaced on a scheduled basis?

☐ Is vacuuming with appropriate equipment used whenever possible rather than blowing or sweeping dust?

☐ Are grinders, saws, and other machines that produce respirable dusts vented to an industrial collector or central exhaust system?

☐ Are all local exhaust ventilation systems designed to provide sufficient airflow and volume for the application, and are ducts not plugged and belts not slipping?

☐ Is PPE provided, used, and maintained wherever required?

☐ Are there written standard operating procedures for the selection and use of respirators where needed?

☐ Are restrooms and washrooms kept clean and sanitary?

☐ Is all water provided for drinking, washing, and cooking potable?

☐ Are all outlets for water that is not suitable for drinking clearly identified?

☐ Are employees' physical capacities assessed before they are assigned to jobs requiring heavy work?

☐ Are employees instructed in the proper manner for lifting heavy objects?

☐ Where heat is a problem, have all fixed work areas been provided with spot cooling or air conditioning?

☐ Are employees screened before assignment to areas of high heat to determine if their health might make them more susceptible to having an adverse reaction?

☐ Are employees working on streets and roadways who are exposed to the hazards of traffic required to wear bright-colored (traffic orange) warning vests?

☐ Are exhaust stacks and air intakes located so that nearby contaminated air will not be recirculated within a building or other enclosed area?

☐ Is equipment producing ultraviolet radiation properly shielded?

☐ Are universal precautions observed where occupational exposure to blood or other potentially infectious materials can occur and in all instances where differentiation of types of body fluids or potentially infectious materials is difficult or impossible?

Flammable and Combustible Materials

☐ Are combustible scrap, debris, and waste materials (oily rags, etc.) stored in covered metal receptacles and promptly removed from the worksite?

☐ Is proper storage practiced to minimize the risk of fire, including spontaneous combustion?

☐ Are approved containers and tanks used to store and handle flammable and combustible liquids?

☐ Are all connections on drums and combustible liquid piping vapor- and liquid-tight?

☐ Are all flammable liquids kept in closed containers when not in use (e.g., parts cleaning tanks, pans, etc.)?

☐ Are bulk drums of flammable liquids grounded and bonded to containers during dispensing?

☐ Do storage rooms for flammable and combustible liquids have explosion-proof lights and mechanical or gravity ventilation?

☐ Is liquefied petroleum gas stored, handled, and used in accordance with safe practices and standards?

☐ Are "NO SMOKING" signs posted on liquefied petroleum gas tanks and in areas where flammable or combustible materials are used or stored?

☐ Are liquefied petroleum storage tanks guarded to prevent damage from vehicles?

☐ Are all solvent wastes and flammable liquids kept in fire-resistant, covered containers until they are removed from the worksite?

☐ Is vacuuming used whenever possible rather than blowing or sweeping combustible dust?

☐ Are firm separators placed between containers of combustibles or flammables that are stacked one upon another to ensure their support and stability?

☐ Are fuel gas cylinders and oxygen cylinders separated by distance and fire-resistant barriers while in storage?

☐ Are fire extinguishers selected and provided for the types of materials in the areas where they are to be used?

☐ Class A—Ordinary combustible material fires.

☐ Class B—Flammable liquid, gas, or grease fires.

☐ Class C—Energized electrical equipment fires.

☐ Are appropriate fire extinguishers mounted within 75 feet (22.86 meters) of outside areas containing flammable liquids and within 10 feet (3.048 meters) of any inside storage area for such materials?

☐ Are extinguishers free from obstructions or blockage?

☐ Are all extinguishers serviced, maintained, and tagged at intervals not to exceed 1 year?

☐ Are all extinguishers fully charged and in their designated places?

☐ Where sprinkler systems are permanently installed, are the nozzle heads so directed or arranged that water will not be sprayed into operating electrical switchboards and equipment?

☐ Are safety cans used for dispensing flammable or combustible liquids at the point of use?

☐ Are all spills of flammable or combustible liquids cleaned up promptly?

☐ Are storage tanks adequately vented to prevent the development of excessive vacuum or pressure as a result of filling, emptying, or atmosphere temperature changes?

☐ Are storage tanks equipped with emergency venting that will relieve excessive internal pressure caused by fire exposure?

☐ Are rules enforced in areas involving storage and use of hazardous materials?

Hazardous Chemical Exposure

☐ Are employees aware of the potential hazards and trained in safe handling practices for situations involving various chemicals stored or used in the workplace, such as acids, bases, caustics, epoxies, phenols, etc.?

☐ Is employee exposure to chemicals kept within acceptable levels?

☐ Are eyewash fountains and safety showers provided in areas where corrosive chemicals are handled?

☐ Are all containers, such as vats, storage tanks, etc., labeled as to their contents, e.g., "CAUSTICS"?

☐ Are all employees required to use personal protective clothing and equipment when handling chemicals (gloves, eye protection, respirators, etc.)?

☐ Are flammable or toxic chemicals kept in closed containers when not in use?

☐ Are chemical piping systems clearly marked as to their content?

☐ Where corrosive liquids are frequently handled in open containers or drawn from storage vessels or pipelines, are adequate means readily available for neutralizing or disposing of spills or overflows and performed properly and safely?

☐ Are standard operating procedures established, and are they being followed when cleaning up chemical spills?

☐ Are respirators stored in a convenient, clean, and sanitary location, and are they adequate for emergencies?

☐ Are employees prohibited from eating in areas where hazardous chemicals are present?

☐ Is PPE used and maintained whenever necessary?

☐ Are there written standard operating procedures for the selection and use of respirators where needed?

☐ If you have a respirator protection program, are your employees instructed on the correct usage and limitations of the respirators?

☐ Are the respirators National Institute for Occupational Safety and Health (NIOSH) approved for this particular application?

☐ Are they regularly inspected, cleaned, sanitized, and maintained?

☐ If hazardous substances are used in your processes, do you have a medical or biological monitoring system in operation?

☐ Are you familiar with the threshold limit values or permissible exposure limits of airborne contaminants and physical agents used in your workplace?

☐ Have appropriate control procedures been instituted for hazardous materials, including safe handling practices and the use of respirators and ventilation systems?

☐ Whenever possible, are hazardous substances handled in properly designed and exhausted booths or similar locations?

☐ Do you use general dilution or local exhaust ventilation systems to control dusts, vapors, gases, fumes, smoke, solvents, or mists that may be generated in your workplace?

☐ Is operational ventilation equipment provided for removal of contaminants from production grinding, buffing, spray painting, and vapor degreasing?

☐ Do employees complain about dizziness, headaches, nausea, irritation, or other factors of discomfort when they use solvents or other chemicals?

☐ Is there a dermatitis problem? Do employees complain about dryness, irritation, or sensitization of the skin?

☐ Have you considered having an industrial hygienist or environmental health specialist evaluate your operation?

☐ If internal combustion engines are used, is carbon monoxide kept within acceptable levels?

☐ Is vacuuming used rather than blowing or sweeping dust whenever possible for cleanup?

☐ Are materials that give off toxic, asphyxiant, suffocating, or anesthetic fumes stored in remote or isolated locations when not in use?

Hazardous Substances Communication

☐ Is there a list of hazardous substances used in your workplace and an MSDS readily available for each hazardous substance used?

☐ Is there a current written exposure control plan for occupational exposure to bloodborne pathogens and other potentially infectious materials, where applicable?

☐ Is there a written hazard communication program dealing with MSDSs, labeling, and employee training?

☐ Is each container for a hazardous substance (i.e., vats, bottles, storage tanks, etc.) labeled with product identity and a hazard warning (communication of the specific health hazards and physical hazards)?

☐ Is there an employee training program for hazardous substances that includes:

- An explanation of what an MSDS is and how to use and obtain one?
- MSDS contents for each hazardous substance or class of substances?
- Explanation of "a right to know"?
- Identification of where an employee can see the written hazard communication program?
- Location of physical and health hazards in particular work areas and the specific protective measures to be used?
- Details of the hazard communication program, including how to use the labeling system and MSDSs?

☐ Does the employee training program on the bloodborne pathogens standard contain the following elements:

- An accessible copy of the standard and an explanation of its contents?
- A general explanation of the epidemiology and symptoms of bloodborne diseases?
- An explanation of the modes of transmission of bloodborne pathogens?
- An explanation of the employer's exposure control plan and the means by which employees can obtain a copy of the written plan?
- An explanation of the appropriate methods for recognizing tasks and the other activities that may involve exposure to blood and other potentially infectious materials?
- An explanation of the use and limitations of methods that will prevent or reduce exposure, including appropriate engineering controls, work practices, and PPE?
- Information on the types, proper use, location, removal, handling, decontamination, and disposal of PPE?
- An explanation of the basis for selection of PPE?
- Information on the hepatitis B vaccine?
- Information on the appropriate actions to take and persons to contact in an emergency involving blood or other potentially infectious materials?
- An explanation of the procedure to follow if an exposure incident occurs, including the methods of reporting the incident and the medical follow-up that will be made available?
- Information on postexposure evaluations and follow-up?
- An explanation of signs, labels, and color coding?

☐ Are employees trained in:

- How to recognize tasks that might result in occupational exposure?
- How to use work practice, engineering controls, and PPE, and their limitations?
- How to obtain information on the types, selection, proper use, location, removal, handling, decontamination, and disposal of PPE?
- Who to contact and what to do in an emergency?

Electrical

☐ Do you require compliance with OSHA standards for all contract electrical work?

☐ Are all employees required to report any obvious hazard to life or property in connection with electrical equipment or lines as soon as possible?

☐ Are employees instructed to make preliminary inspections or appropriate tests to determine conditions before starting work on electrical equipment or lines?

☐ When electrical equipment or lines are to be serviced, maintained, or adjusted, are necessary switches opened, locked out, or tagged whenever possible?

☐ Are portable electrical tools and equipment grounded or of the double insulated type?

☐ Are electrical appliances such as vacuum cleaners, polishers, vending machines, etc., grounded?

☐ Do extension cords have a grounding conductor?

☐ Are multiple plug adaptors prohibited?

☐ Are ground fault circuit interrupters installed on each temporary 15- or 20-ampere, 120-volt alternating current (AC) circuit at locations where construction, demolition, modifications, alterations, or excavations are being performed?

☐ Are all temporary circuits protected by suitable disconnecting switches or plug connectors at the junction with permanent wiring?

☐ Do you have electrical installations in hazardous dust or vapor areas? If so, do they meet the National Electrical Code (NEC) for hazardous locations?

☐ Are exposed wiring and cords with frayed or deteriorated insulation repaired or replaced promptly?

☐ Are flexible cords and cables free of splices or taps?

☐ Are clamps or other securing means provided on flexible cords or cables at plugs, receptacles, tools, equipment, etc., and is the cord jacket securely held in place?

☐ Are all cord, cable, and raceway connections intact and secure?

☐ In wet or damp locations, are electrical tools and equipment appropriate for the use or location or otherwise protected?

☐ Is the location of electrical power lines and cables (overhead, underground, under floor, other side of walls, etc.) determined before digging, drilling, or similar work is begun?

☐ Are metal measuring tapes, ropes, handlines, or similar devices with metallic thread woven into the fabric prohibited where they could come in contact with energized parts of equipment or circuit conductors?

☐ Is the use of metal ladders prohibited where the ladder or the person using the ladder could come in contact with energized parts of equipment, fixtures, or circuit conductors?

☐ Are all disconnecting switches and circuit breakers labeled to indicate their use or equipment served?

☐ Are disconnecting means always opened before fuses are replaced?

☐ Do all interior wiring systems include provisions for grounding metal parts of electrical raceways, equipment, and enclosures?

☐ Are all electrical raceways and enclosures securely fastened in place?

☐ Are all energized parts of electrical circuits and equipment guarded against accidental contact by approved cabinets or enclosures?

☐ Is sufficient access and working space provided and maintained around all electrical equipment to permit ready and safe operations and maintenance?

☐ Are all unused openings (including conduit knockouts) in electrical enclosures and fittings closed with appropriate covers, plugs, or plates?

☐ Are electrical enclosures such as switches, receptacles, junction boxes, etc., provided with tight-fitting covers or plates?

☐ Are disconnecting switches for electrical motors in excess of 2 horsepower able to open the circuit when the motor is stalled without exploding? (Switches must be horsepower rated equal to or in excess of the motor rating.)

☐ Is low-voltage protection provided in the control device of motors driving machines or equipment that could cause injury from inadvertent starting?

☐ Is each motor disconnecting switch or circuit breaker located within sight of the motor control device?

☐ Is each motor located within sight of its controller, or is the controller disconnecting means able to be locked open, or is a separate disconnecting means installed in the circuit within sight of the motor?

☐ Is the controller for each motor that exceeds 2 horsepower rated equal to or above the rating of the motor it serves?

☐ Are employees who regularly work on or around energized electrical equipment or lines instructed in cardiopulmonary resuscitation (CPR)?

☐ Are employees prohibited from working alone on energized lines or equipment over 600 volts?

Noise

☐ Are there areas in the workplace where continuous noise levels exceed 85 decibels?

☐ Is there an ongoing preventive health program to educate employees in safe levels of noise, exposures, effects of noise on their health, and the use of personal protection?

☐ Have work areas where noise levels make voice communication between employees difficult been identified and posted?

☐ Are noise levels measured with a sound level meter or an octave band analyzer, and are records being kept?

☐ Have engineering controls been used to reduce excessive noise levels? Where engineering controls are determined to be infeasible, are administrative controls (i.e., worker rotation) being used to minimize individual employee exposure to noise?

☐ Is approved hearing protective equipment (noise-attenuating devices) available to every employee working in noisy areas?

☐ Have you tried isolating noisy machinery from the rest of your operation?

☐ If you use ear protectors, are employees properly fitted and instructed in their use?

☐ Are employees in high-noise areas given periodic audiometric testing to ensure that you have an effective hearing protection system?

Fueling

☐ Are employees prohibited from fueling an internal combustion engine with a flammable liquid while the engine is running?

☐ Are fueling operations performed to minimize spillage?

☐ When spillage occurs during fueling operations, is the spilled fuel washed away completely, evaporated, or are other measures taken to control vapors before restarting the engine?

☐ Are fuel tank caps replaced and secured before starting the engine?

☐ In fueling operations, is there always metal contact between the container and the fuel tank?

☐ Are fueling hoses designed to handle the specific type of fuel?

☐ Are employees prohibited from handling or transferring gasoline in open containers?

☐ Are open lights, open flames, sparking, or arcing equipment prohibited near fueling or transfer of fuel operations?

☐ Is smoking prohibited in the vicinity of fueling operations?

☐ Are fueling operations prohibited in buildings or other enclosed areas that are not specifically ventilated for this purpose?

☐ Where fueling or transfer of fuel is done through a gravity flow system, are the nozzles self-closing?

Identification of Piping Systems

☐ When nonpotable water is piped through a facility, are outlets or taps posted to alert employees that the water is unsafe and not to be used for drinking, washing, or other personal use?

☐ When hazardous substances are transported through aboveground piping, is each pipeline identified at points where confusion could introduce hazards to employees?

☐ When pipelines are identified by color-painted bands or tapes, are the bands or tapes located at reasonable intervals and at each outlet, valve, or connection, and are all visible parts of the line so identified?

☐ When pipelines are identified by color, is the color code posted at all locations where confusion could introduce hazards to employees?

☐ When the contents of pipelines are identified by name or name abbreviation, is the information readily visible on the pipe near each valve or outlet?

☐ When pipelines carrying hazardous substances are identified by tags, are the tags constructed of durable materials, the message printed clearly and permanently, and are tags installed at each valve or outlet?

☐ When pipelines are heated by electricity, steam, or other external source, are suitable warning signs or tags placed at unions, valves, or other serviceable parts of the system?

Materials Handling

☐ Is there safe clearance for equipment through aisles and doorways?

☐ Are aisleways permanently marked and kept clear to allow unhindered passage?

☐ Are motorized vehicles and mechanized equipment inspected daily or prior to use?

☐ Are vehicles shut off and brakes set prior to loading or unloading?

☐ Are containers of liquid combustibles or flammables, when stacked while being moved, always protected by dunnage (packing material) sufficient to provide stability?

☐ Are dock boards (bridge plates) used when loading or unloading operations are taking place between vehicles and docks?

☐ Are trucks and trailers secured from movement during loading and unloading operations?

☐ Are dock plates and loading ramps constructed and maintained with sufficient strength to support imposed loading?

☐ Are hand trucks maintained in safe operating condition?

☐ Are chutes equipped with sideboards of sufficient height to prevent the materials being handled from falling off?

☐ Are chutes and gravity roller sections firmly placed or secured to prevent displacement?

☐ Are provisions made to brake the movement of the handled materials at the delivery end of rollers or chutes?

☐ Are pallets usually inspected before being loaded or moved?

☐ Are safety latches and other devices being used to prevent slippage of materials off of hoisting hooks?

☐ Are securing chains, ropes, chockers, or slings adequate for the job?

☐ Are provisions made to ensure that no one is below when hoisting material or equipment?

☐ Are MSDSs available to employees handling hazardous substances?

Transporting Employees and Materials

☐ Do employees who operate vehicles on public thoroughfares have valid operator's licenses?

☐ When seven or more employees are regularly transported in a van, bus, or truck, is the operator's license appropriate for the class of vehicle being driven, and are there enough seats?

☐ Are vehicles used to transport employees equipped with lamps, brakes, horns, mirrors, windshields, and turn signals, and are they in good repair?

☐ Are transport vehicles provided with handrails, steps, stirrups, or similar devices, placed and arranged to allow employees to safely mount or dismount?

☐ Are employee transport vehicles equipped at all times with at least two reflective-type flares?

☐ Is a fully charged fire extinguisher, in good condition, with at least a 4 B:C rating maintained in each employee transport vehicle?

☐ When cutting tools or tools with sharp edges are carried in passenger compartments of employee transport vehicles, are they placed in closed boxes or containers that are secured in place?

☐ Are employees prohibited from riding on top of any load that could shift, topple, or otherwise become unstable?

Control of Harmful Substances by Ventilation

☐ Is the volume and velocity of air in each exhaust system sufficient to gather the dusts, fumes, mists, vapors, or gases to be controlled, and to convey them to a suitable point of disposal?

☐ Are exhaust inlets, ducts, and plenums designed, constructed, and supported to prevent collapse or failure of any part of the system?

☐ Are clean-out ports or doors provided at intervals not to exceed 12 feet (3.6576 meters) in all horizontal runs of exhaust ducts?

☐ Where two or more different operations are being controlled through the same exhaust system, could the combination of substances involved create a fire, explosion, or chemical reaction hazard in the duct?

☐ Is adequate makeup air provided to areas where exhaust systems are operating?

☐ Is the source point for makeup air located so that only clean, fresh air, free of contaminants, will enter the work environment?

☐ Where two or more ventilation systems serve a work area, is their operation such that one will not offset the functions of the other?

Sanitizing Equipment and Clothing

☐ Is required personal protective clothing or equipment able to be cleaned and disinfected easily?

☐ Are employees prohibited from interchanging personal protective clothing or equipment, unless it has been properly cleaned?

☐ Are machines and equipment that process, handle, or apply materials that could injure employees cleaned or decontaminated before being overhauled or placed in storage?

☐ Are employees prohibited from smoking or eating in any area where contaminants are present that could be injurious if ingested?

☐ When employees are required to change from street clothing into protective clothing, is a clean change room with a separate storage facility for street and protective clothing provided?

☐ Are employees required to shower and wash their hair as soon as possible after a known contact with a carcinogen has occurred?

☐ When equipment, materials, or other items are taken into or removed from a carcinogen-regulated area, is it done in a manner that will not contaminate nonregulated areas or the external environment?

Tire Inflation

☐ Where tires are mounted or inflated on drop center wheels or on wheels with split rims or retainer rings, is a safe practice procedure posted and enforced?

☐ Does each tire inflation hose have a clip-on chuck with at least 2.54 inches (6.45 centimeters) of hose between the chuck and an in-line hand valve and gauge?

☐ Does the tire inflation control valve automatically shut off the airflow when the valve is released?

☐ Is a tire restraining device such as a cage, rack, or other effective means used while inflating tires mounted on split rims or rims using retainer rings?

☐ Are employees prohibited from standing directly over or in front of a tire while it is being inflated?

ASSISTANCE IN SAFETY AND HEALTH FOR SMALL BUSINESSES

OSHA ASSISTANCE

OSHA's Office of Small Business Assistance

OSHA created the Office of Small Business Assistance to help small business employers understand their safety and health obligations, access compliance information, provide guidance on regulatory standards, and to educate them about cost-effective means for ensuring the safety and health of worksites.

OSHA's Office of Small Business Assistance can be contacted by telephone at (202) 693-2220 or by writing to the U.S. Department of Labor, 200 Constitution Avenue, NW, Room N-3700, Washington, DC 20210.

On-Site Consultation

Using the free and confidential on-site consultation service largely funded by the federal OSHA, employers can find out about potential hazards at their worksites, improve their occupational safety and health management systems, and even qualify for a 1-year exemption from routine OSHA inspections.

The service is delivered at your workplace by state governments using well-trained professional staff. Most consultations take place on-site, though limited services away from the worksite are available.

Primarily targeted for smaller businesses, this safety and health consultation program is completely separate from OSHA's enforcement efforts. It is also confidential. No inspections are triggered by using the consultation program, and no citations are issued or penalties proposed.

Your name, your firm's name, and any information you provide about your workplace, plus any unsafe or unhealthful working conditions that the consultant uncovers, will not routinely be reported to the OSHA enforcement staff.

Your only obligation will be to commit to correcting serious job safety and health hazards discovered—a commitment that you are expected to make prior to the actual consultation visit. If hazards are discovered, the consultant will work with you to ensure they are corrected in a reasonable timeframe agreed upon by all parties.

Getting started. Since consultation is a voluntary activity, you must request it. Your telephone call or letter sets the consulting machinery in motion. The consultant will discuss your specific needs and set up a visit date based on the priority assigned to your request, your work schedule, and the time needed for the consultant to prepare adequately to serve you. OSHA encourages a complete review of your firm's safety and health situation; however, if you wish, you may limit the visit to one or more specific problems.

Opening conference. When the consultant arrives at your worksite for the scheduled visit, he or she will first meet with you in an opening conference to briefly review the consultant's role and the obligations you incur as an employer.

Walk-through. Together, you and the consultant will examine conditions in your workplace. OSHA strongly encourages maximum employee participation in the walk-through. Better-informed and alert employees can help you identify and correct potential injury and illness hazards in your workplace. Talking with employees during the walk-through helps the consultant identify and judge the nature and extent of specific hazards.

The consultant will study your entire workplace, or only those specific operations you designate, and discuss applicable OSHA standards. The consultant also will point out other safety or health risks that might not be cited under OSHA standards, but which nevertheless may pose safety or health risks to your employees. He or she may suggest and even provide measures such as self-inspection and safety and health training that you and your employees can apply to prevent future hazardous situations.

A comprehensive consultation also includes: (1) appraisal of all mechanical and environmental hazards and physical work practices, (2) appraisal of the present job safety and health program or help in establishing one, (3) a conference with management on findings, (4) a written report of recommendations and agreements, and (5) training and assistance with implementing recommendations.

Closing conference. The consultant will then review detailed findings with you in a closing conference. You will learn not only what you need to improve, but what you are doing right as well. At that time you can discuss problems, possible solutions, and abatement periods to eliminate or control any serious hazards identified during the walk-through.

In rare instances, the consultant may find an imminent danger situation during the walk-through. In that case, you must take immediate action to protect employees. In certain other situations—those that would be judged a serious violation under OSHA criteria—you and the consultant must develop and agree to a reasonable plan and schedule to eliminate or control that hazard. The consultant will offer general approaches and options to you. He or she may also suggest other sources for technical help.

Abatement and follow-through. Following the closing conference, the consultant will send you a detailed written report explaining the findings and confirming any abatement periods agreed upon. The consultant may also contact you from time to time to check your progress. You, of course, may always contact him or her for assistance.

Ultimately, OSHA does require hazard abatement so that each consultation visit achieves its objective-effective employee protection. If you fail to eliminate or control identified serious hazards (or an imminent danger) according to the plan and within the limits agreed upon or an agreed-upon extension, the situation must be referred from consultation to an OSHA enforcement office for appropriate action. This type of referral is extremely rare.

Benefits. Knowledge of your workplace hazards and ways to eliminate them can only improve your own operations—and the management of your firm. You will get professional advice and assistance on the correction of workplace hazards and benefit from on-site training and assistance provided.

The consultant can help you establish or strengthen an employee safety and health program, making safety and health activities routine rather than crisis-oriented responses.

In many states, employers may participate in OSHA's Safety and Health Achievement Recognition Program (SHARP). This program is designed to provide incentives and support to smaller, high-hazard employers to develop, implement, and continuously improve effective safety and health programs at their worksite(s). SHARP provides recognition of employers who have demonstrated exemplary achievements in workplace safety and health, beginning with a comprehensive safety and health consultation visit, correction of all workplace safety and health hazards, adoption and implementation of effective safety and health management systems, and agreement to request further consultative visits if major changes in working conditions or processes occur that may introduce new hazards. Employers who meet these specific SHARP requirements may be removed from OSHA's programmed inspection list for 1 year.

The on-site consultants will:

- Help you recognize hazards in your workplace
- Suggest general approaches or options for solving a safety or health problem
- Identify kinds of help available if you need further assistance
- Provide you with a written report summarizing findings
- Assist you in developing or maintaining an effective safety and health program
- Provide training and education for you and your employees
- Recommend you for a 1-year exclusion from OSHA programmed inspections, once program criteria are met

The on-site consultants will not:

- Issue citations or propose penalties for violations of OSHA standards
- Report possible violations to OSHA enforcement staff
- Guarantee that your workplace will pass an OSHA inspection

For a list of consultation projects in each state, see the OSHA website.[*]

---

[*] From the small business safety publication located at www.osha.gov.

Appendix H: Safety Auditing

Throughout our lives, we are evaluated and graded on our performance; e.g., we received a grade of A in math in third grade. However, traditionally in safety and health, when the boss asked you how the safety and health program was doing, the only objective data that could be provided were the reactive numerical values related to injuries and illnesses. Do injury and illness data truly reflect the status of the proactive safety and health effort?

In order to know your objective program status, program effectiveness, and program deficiencies, it is imperative that safety and health professionals develop a safety and health audit system that can provide objective data as to the status of each and every element of each and every compliance program, as well as all other aspects of the overall safety and health program. Although there are a number of different scoring or grading systems, it is important that the overall safety and health program be carefully and accurately evaluated on at least an annual basis.

SAFETY AND HEALTH PROGRAM AUDITS AND REVIEWS

AUDIT TRAIL

Initiating an audit trail for evaluating a safety and health program is a way to test the effectiveness of written or informal programs. Depending on the amount of resources and time you want to devote, the process can be as simple as taking several of the incidents you may see listed on a company's annual OSHA 200 (annual summary of occupational injuries and illnesses) and tracking back through applicable company reports or programs. By taking this tactic, you can get a pretty good idea whether an effective safety and health program has been implemented.

Take the example of two eye injuries that were entered on one company's OSHA 200. The first step could be pulling the 101s or First Reports of Accidents or Illness for the two eye injuries. Full evaluation of the reported information includes first checking on proper recording, and then evaluating background information as to why the eye injury occurred. The company had listed the cause of the injury as employee failure to wear eye protection.

ROOT CAUSE ANALYSIS

This is only the starting point for the audit trail. Now, the real search for root causes and the deeper evaluation of the company's programs can begin. Employee interviews revealed that one employee felt that the goggles provided did not fit. The other employee complained of the goggles fogging up as the reason that the goggles were not being worn at the time of the injury. An interview with the manager revealed that the manager was reluctant to initiate the company disciplinary policy for the two employees who were excellent workers. Typical of most companies, the accident report blamed the victim. The company had well-written safety and health programs, performed recordkeeping accurately, and had trained its employees thoroughly. What was missing from its safety and health program?

IMPACT OF TOTAL QUALITY MANAGEMENT (TQM) PRINCIPLES ON SAFETY AND HEALTH PROGRAMS

If this company had applied basic TQM principles, it would have encouraged full evaluation of the sources of nonconformance (not wearing the goggles), rather than turning to discipline. If the manager had questioned all of the employees who perform the job that had resulted in two eye injuries, she would have found that 75% of the time employees indicated that they did not wear their goggles because they forgot them in their locker, 10% of the time employees felt that the goggles did not fit or slipped off, and 15% of the time employees felt that goggle fogging was a big problem. By addressing each of these issues separately, incidences of nonconformance could be significantly reduced. The manager could purchase retainer clips that attached the goggles directly to the hard hats. Changing style of goggles could eliminate the problem of lenses fogging. The manager also could purchase an extra supply so that any goggle damaged by chemicals could be immediately changed as needed.

If only the eye injury cases had been evaluated, a significant reason for not using the goggles (representing significant risk) would not have been addressed. In other words, the 75% of the employees who did not wear their goggles because they had left them in their lockers would not have been addressed.

If the employees had participated in the accident evaluation, additional input might have been garnered at the time of the first incident, preventing the second incident. Also, there was no discussion of the injury and illness entries or near misses each month during the safety and health committee meeting, indicating a failure to use the safety and health committee to address real, practical, and soluble problems in the workplace. Had the safety and health committee addressed the root causes during their meeting, they might also have been able to address all of the instances of nonconformance before the second injury.

AUDIT TRAIL STEPS

After the first injury the following sequence of events should have taken place:

- Perform accident analysis and appropriate recordkeeping.
- Review accident report during the safety and health committee meeting for further input, and determine corrective actions.
- Determine whether any near misses had occurred for this job, and if so, why they had not been reported.
- Review job safety hazard analysis (JSA) for the job in question, and revise as necessary.
- Reevaluate disciplinary/incentive programs and how they impact true employee participation and reporting of near misses. Modify if they are disincentives.

- Retrain the employees and managers accordingly.
- Perform an analysis of your conformance with identified critical safety behaviors to evaluate the effectiveness of corrective actions.

ADDITIONAL POSITIVE OUTCOMES

If the previous actions had been completed, a number of positive results could have occurred. First, by addressing the root causes, the manager would have demonstrated her commitment to preventing eye injuries. Employee participation in the safety and health committee, accident analysis, job safety hazard analysis, and retraining employees could have heightened employee knowledge and sensitivity to this critical safety behavior, and increased employee morale and feeling of being part of a team. As a result of goggle use becoming routine, eye injuries might have been reduced significantly or eliminated entirely.

Although reviewing OSHA 200 entries is a good tool for auditing a safety and health program, it is only one of many tools that can be used. Such an approach takes only a retroactive look at programs once a problem has occurred. Ideally, a company should try to be proactive in its approach.

What this audit approach does, however, is to look to see if the company's program is a "walking/working" program, with linkages that interconnect with other parts of the program. In effect, if there is an accident or a near miss, this process can be followed to see if there was the appropriate ripple effect. Would the near miss be reported? Once reported, would the JSA be reviewed, or would employees be retrained? Making an accident report, because it is expected, but not seeing evidence of the ripple effect, such as changing a JSA or retraining employees or purchasing new equipment, may be evidence that a safety and health program is a "paper program" and not a vital working document.

---

* www.osha.gov.

Appendix I: Action Planning Document

After the development of a compliance program and the necessary approvals, the next step for most safety and health professionals is the implementation of the program within the operations. Although there is no one "right way" to develop an action plan, it is essential that safety and health professionals develop a method through which to keep track of each and every activity to ensure a timely and accurate implementation. Many safety and health professionals develop action planning documents in a spreadsheet format or within other software programs on their computers for easy access. The basic questions of who, what, when, where, and how should be addressed. Additionally, an action planning document can provide accurate information to keep the program implementation on track and provide specific information as to who within the organization is responsible for what part of the implementation process. Below is an example of a basic action planning document.

EXAMPLE ACTION PLANNING DOCUMENT

Compliance Program—Lockout and Tagout (LOTO)

Program Element	Initiation Date	Completion Date	Location of Activity	Responsible Party	Verification of Completion
1. Purchase Locks	1/1/13	1/21/13	Purchasing	Tom S.	1/22/13
2. LOTO Training	1/15/13	2/25/13	Training Room	Ron D.	3/1/13

Index